THE SIMON AND SCHUSTER
STEP-BY-STEP ENCYCLOPEDIA
OF PRACTICAL GARDENING
Published in cooperation with the Royal Horticultural Society

Garden Pests and Diseases

by Audrey Brooks and Andrew Halstead

Editor-in-chief Christopher Brickell
Technical editor Kenneth A. Beckett

Editor Martin Elliott
Art editor Tony Spalding
Editorial assistant Helen Buttery
Designers Janet Baker, Winnie Malcolm,
  Michael Nawrocki, Alan Wood
Picture researcher Robert Wyburn
Executive editor Chris Foulkes

Published by Simon and Schuster
A Division of Gulf & Western Corporation
Simon & Schuster Building
Rockefeller Center,
1230 Avenue of the Americas,
New York, New York 10020

ISBN 0 671 42254 5
Library of Congress Catalog Card Number: 80–5863

Garden Pests and Diseases was edited and designed by
Mitchell Beazley Publishers Limited, Mill House,
87–89 Shaftesbury Avenue, London W1V 7AD

Typesetting by Tradespools Ltd, Frome, Somerset
Origination by Culver Graphics Ltd,
High Wycombe, Buckinghamshire
Printed in Spain by Printer industria gráfica sa.,
Sant Vicenç dels Horts, Barcelona
Depósito Legal B—16691—1980

Contents

Introduction 1

The range of plants grown by amateur gardeners is extremely wide and, correspondingly, there are many pests, diseases and disorders that can occur. In this book the various troubles that may affect a particular plant are listed by symptom, for example, Leaves discolored, followed by a description of the possible causes and the correct control measures in each case. This enables the gardener to avoid the hit and miss methods that are too often used by amateurs. Some have been known to use first one insecticide and then another, and then perhaps several different fungicides, to try to cure an unhealthy plant, when the only treatment needed was to lift it and replant it more carefully. If the correct control measures are used at the onset of symptoms, not only will time, labor and money be saved, but there will also be less risk of disturbing the balance of nature by the misuse of chemicals.

This book has been extensively rewritten from the original British edition in order to adapt it to the needs of the American gardener. In particular, many of the pesticides recommended in the United Kingdom have been changed to conform to the requirements of United States laws.

The authority for recommending or banning pesticides is vested in the Federal Government's EPA (Environmental Protection Agency). Most States have also formed their own regulatory agencies such as DEP (Department of Environmental Protection) which have put even more drastic regulations into effect. As a result some pesticides for control of a specific pest are approved by one State and banned in an adjoining State.

Almost weekly, new regulations to approve or ban a given pesticide are forthcoming from EPA and the State regulatory agencies. For this reason, readers are advised to check with their own State Agricultural Experiment Stations for control of the diseases and insects in their area.

Pests

The term pest encompasses all animals that damage plants, including vertebrates such as birds and mammals, as well as the smaller animals such as insects, mites, mollusks and eelworms. If the pest can be seen on the plant then identification is fairly easy, but often, particularly with microscopic pests such as eelworms and mites, it is necessary to identify the pest from the symptoms it has caused. These symptoms vary according to the plant, and the pest that is involved, but they are generally related to the way in which the pest feeds.

Some insects and mites have needle-like mouthparts that are used to suck sap from plants. They attack mainly leaves and stems and as they feed, certain of these pests, particularly aphids and capsid bugs, inject chemicals into the foliage, causing leaf curling and distorted growth. Other sap-feeding pests, such as eelworms, live inside the stems, leaves or roots of their host plants where they take sap from plant cells. Many of these sap-feeding creatures can carry virus particles on their mouthparts, and transmit them when they move from an infected plant to a healthy one. Another group of pests have biting or rasping mouthparts, which are used to bite holes in the foliage, stems and roots. They include caterpillars, sawfly larvae, fly maggots, beetles, earwigs, slugs and snails.

A further symptom caused by certain insects, mites and eelworms is the production of abnormal growths known as galls, in which the causal animal lives and feeds. These galls sometimes give the leaves and stems a bizarre appearance, but in many cases no real harm is done to the host plant and control measures are not usually necessary.

Diseases

Diseases are caused by parasitic organisms called pathogens, of which there are several different types. The most important pathogens are the fungi. These are plants that contain no chlorophyll (green color) and so are unable to make their own carbohydrates. Instead, they consume food already manufactured by green plants. The vast majority of fungi live on dead and decaying matter, but a few obtain their food from living plants; it is this latter group that are responsible for fungal diseases.

Most fungi are composed of microscopic threads, and so can only be identified by the symptoms and not by the organism. On some species, however, these threads join together to form large structures such as toadstools and bracket fungi; in these cases identification becomes easier.

Bacteria are much smaller than fungi, being only just visible with an ordinary microscope. They cannot break down the protective outer layer that plants possess as can most pathogenic fungi, so infection takes place through wounds or natural openings. These organisms can only be identified by experts, and gardeners have to recognize the disease by the symptoms produced.

More important as pathogens than bacteria or actinomycetes are the viruses. They are extremely small, being no more than a few hundred-thousandths of an inch long, and have more in common with chemicals than with living organisms. Nevertheless, they are able to multiply within the cells of plants. Most viruses are introduced into plants by pests, particularly aphids or eelworms. A few viruses are seed-borne or pollen-borne. They can also enter a plant through wounds produced when plants are handled roughly and by physical contact between a diseased and a healthy plant. Once infection has occurred nearly every cell of a diseased plant becomes infected. There is, therefore, no simple method by which a gardener can cure such a plant, and it should be destroyed.

Cultural control of pests and diseases

The most important means of combating pests and diseases is by good cultural practice. Thorough preparation of the soil before planting or sowing will ensure that plants are given a good start. Fertilizers and liming may be needed, depending on local soil conditions and the requirements of the plants. Most plants will benefit from a regular supply of water, especially if dry spells coincide with critical stages in the plant's growth, such as germination, bud formation, flowering and fruiting. Mulching the soil surface with well rotted manure or some similar layer of organic matter will help to retain soil moisture, as well as providing some nutrients and suppressing weeds.

Seeds should be sown at the correct time, taking into account local conditions, since if they are sown prematurely or too late the plants may fail to thrive. Similarly germination will be poor if the seeds are sown at the wrong depth. The rate of sowing is also important since sowing too densely will result in seedlings that are thin, weak and susceptible to disease because they are competing with each other for water, nutrients and space.

With plants raised in containers the time of planting out is less important, but it is necessary to make sure that they are planted with the soil surface within the container flush with the level of the garden soil. On bare-rooted trees and shrubs there is a change of color, or soil mark, at the base of the stem which indicates the correct level for planting.

When planting any type of plant, but particularly trees and shrubs, a hole large enough to accommodate the roots should be prepared. However, if the soil is very heavy and subject to waterlogging, the hole should not be filled with a lighter material such as peat or compost because it will then act as a sump into which water from the surrounding heavier soil will flow, making matters worse. Any treatments to lighten heavy soil should be carried out over a wide area and not just in the planting holes. When digging holes in clay soil do not use a spade or trowel because such implements will compress the sides of the holes and the plant's roots may not be able to penetrate outwards.

The roots of most plants need to be well spread out when planting. If the roots are markedly one-sided or entwined around each other and cannot be spread out, soak the roots in a bucket of water for a few days until they are more pliable. Similarly, if a plant is very pot-bound, whether it be a container-grown plant to be planted out in the garden or a pot plant to be potted on, the root system should be held under running water for a short period or dipped in a bucket of water. This allows some of the original compost to be dislodged, and a few roots can be teased out at intervals around the soil ball.

The choice of plants is important in the control of pests and diseases, and local advice should be sought about which plants are likely to do well under the local soil and climatic conditions. Seeds, plants, bulbs and other planting material should be obtained

Introduction 2

from reputable suppliers. Cheap lots from food markets or mail-order concerns may be carrying pests and diseases that can take years to eliminate from a garden. There are some plant varieties that are resistant or have a degree of tolerance to certain pests and diseases. These are mentioned later in the book under the appropriate pest or disease.

Rotation

Growing the same, or closely related, plants on one site in successive years should be avoided as far as possible since soil-borne pests and diseases may build up; this applies as much to ornamental bedding plants as to vegetables. Weeds should be kept under control since, as well as competing with cultivated plants for water and nutrients, many are hosts of pests and diseases. Their presence can, in some cases, nullify the benefits of plant rotation.

Hygiene

Plants that have been badly attacked by pests and diseases should not be left in the garden to infect other plants. Pest-infested plants can generally be disposed of by burying them in a compost heap, but infested roots should always be burned. Most diseased material is best disposed of by burning. In addition, reduce the possibility of pests and diseases carrying over from one year to the next by collecting fallen leaves and fruits, removing dead shoots and cleaning out hedge bottoms. All such debris should be burned.

Chemical control

Gardeners who wish to keep pests and diseases to a minimum will need to use some chemicals for control. Before taking any control measures it is most important to identify correctly the pest, disease or disorder so that the appropriate treatment can be applied. Garden chemicals must be treated with respect at all times, since incorrect use may harm the user or damage plants. **The manufacturer's instructions must be read and followed. Chemicals should be stored in a cool dark place away from foodstuffs, if possible in a locked cupboard where children and pets cannot reach them.**

When diluting chemicals for use it is advisable to wear rubber gloves and to mix no more than will be required for the job. Undiluted chemicals can be stored for years but, once diluted, they start to break down and will be ineffective if used later. When applying insecticides, fungicides and foliar feeds, use equipment that is kept solely for this purpose and has not previously been used for applying a weedkiller. After application, thoroughly wash the sprayer, rubber gloves and other equipment.

Avoid spraying or dusting in windy weather since the chemicals may drift in the wind. The best conditions for spraying are when it is dry, calm and frost-free. There are few days when it is completely calm, however, but spraying can be done when there is a light breeze provided some arrangements are made to screen adjacent plants, for instance with plastic sheeting, if the spray is likely to damage them. The spray should be applied from all sides of the plant to give as good a coverage as possible. The objective when spraying with chemicals that act by contact is to obtain an even cover on both the upper and lower leaf surfaces. When using such chemicals, spray to run-off but no more than this, otherwise the material is wasted.

Never use insecticides on plants that are in flower because this will endanger bees. If large-scale spraying with insecticides is to be carried out, for example in an orchard, always notify local beekeepers so that hives can be closed. Fish and other forms of water life are extremely sensitive to pesticides, so avoid contaminating ponds and waterways with spray drift or old pesticide cans.

Formulations

Garden chemicals are sold in a number of different formulations. The majority are available as liquid concentrates or wettable powders which are mixed with water for application as a spray. Insecticides and fungicides usually contain materials known as spreader-stickers which facilitate contact between the spray and the sprayed surface and assist in the even distribution of the spray over the target. Certain other chemicals, however, such as magnesium sulfate, do not contain spreaders and these have to be added to the solution as indicated in the section on physiological disorders, pages 5–7.

Spraying is usually the most effective and economical method of applying a chemical, but there are situations when another method of application may be more appropriate. Aerosols are comparatively expensive but they require no mixing and are convenient for treating a small number of plants. Dusts also require no mixing and are useful for treating low-growing plants and those that are being attacked by soil-borne pests and diseases. They are less effective than sprays where it is necessary to have a good deposit of the chemical on the undersides of the leaves. Some insecticides are formulated as granules which slowly release the chemical into the soil and so remain active against soil pests for a longer period than if the same chemical were applied as a spray or dust. Smoke formulations, which are very quick and easy to apply, are available for treating certain greenhouse pests and diseases.

Most insecticides and some fungicides have a contact action: the pest or fungus is only killed if it is hit by the chemical when spraying. It is therefore necessary to apply the chemical as thoroughly as possible on both sides of the foliage. Some chemicals are systemic, which means that they are absorbed into the plant's tissues and move in the sap-stream to parts of the plant that have not been sprayed directly. Systemic insecticides are particularly effective against sap-feeding pests, but are less effective against pests with chewing mouthparts; these are better controlled by contact insecticides.

Systemic fungicides such as benomyl and thiophanate-methyl are not truly systemic since they are only translocated short distances inside plants, and so have to be used almost as frequently as other types of fungicides. Most fungicides used by gardeners act as protectants in that they reduce infections by preventing the germination of spores. It is important, therefore, that they are applied before any symptoms are seen or at the very first signs of trouble.

The insecticides and fungicides mentioned in this book are the active ingredients rather than the brand names of garden chemicals. In some cases, the active ingredient is included in the brand name. Alternatively, suitable brands can be found by reading the small print on the labels, since this will always include the product's active ingredient.

Biological control

This method of control utilizes natural enemies to eliminate certain pests. With garden plants there are too many uncontrollable factors for this approach to have much success, but there are several very effective biological controls for certain pests of greenhouse and house plants. There is a predatory mite, *Phytoseiulus persimilis*, for greenhouse red spider mite; a ladybird predator, *Cryptolaemus montrouzeri*, for mealybugs; a parasitic wasp, *Encarsia formosa*, for greenhouse whitefly; and for caterpillars a bacterium, *Bacillus thuringiensis*. All these controls are available from specialist producers who can be located through their advertisements in gardening magazines.

There are certain conditions that are necessary for biological control to be effective. The controls mentioned above are very host-specific and will soon starve in the absence of their prey. It is therefore necessary to wait until the pest is present before introducing the appropriate biological control. The introduction should, however, be made before the pest becomes very numerous, since predators and parasites need time to increase their numbers before they can give effective control. Also, if they are to breed faster than the pests, then daytime temperatures of 21°C/70°F or more and a high light intensity are required. For this reason biological control is only effective between late April and September.

Beneficial insects and mites are very susceptible to most insecticides, and the only one that can be used safely in their presence is pirimicarb, which controls aphids only. Biological control has a number of advantages over chemical control in that there are no insecticide residues on food plants, no problems with spray-sensitive plants and insecticide-resistant pests are fully susceptible to biological controls. Lastly, there is less work involved since once the predators or parasites are established they will look after themselves.

Glossary

Abort To cease growth in the early stages of formation.

Acid Applied to soil with a pH of below 7.

Adventitious roots Roots that develop from stems.

Alkaline Applied to soil with a pH over 7.

Anther The pollen-producing structure at the top of the stamen of a flower.

Basal At the lowest part of the plant or one of its organs.

Bedding plant A plant used for temporary garden display.

Blindness A condition in which a shoot or bud fails to develop fully and aborts.

Bloom Either a blossom, or a natural, white powdery or waxy sheen on many fruits and leaves, or an abnormal white powdery coating of fungus spores on galled leaves.

Bolting Producing flowers and seed prematurely.

Breathing pore A small pore in the surface of woody stems through which gases enter and leave the internal tissues.

Bulb An underground storage organ that consists of layers of swollen fleshy leaves or leaf bases, which enclose the following year's growth bud.

Canker A sharply defined diseased area on a woody stem, which often has malformed bark.

Caterpillar see Larva.

Chelated Describes a special formulation of plant nutrients, which will remain available in alkaline soils.

Chlorophyll The green pigment present in most plants, by means of which they manufacture carbohydrates.

Chlorosis The loss, or poor production, of chlorophyll, causing yellowing of leaves or parts of leaves, or their complete whitening.

Conifer A plant that bears its seeds in cones.

Contact insecticide An insecticide that kills pests with which it comes into contact.

Contractile roots Roots of bulbs and corms that contract in length, thereby pulling the organ deeper into the soil.

Corm A solid, swollen stem-base, resembling a bulb, that acts as a storage organ.

Cotyledon A seed leaf; usually the first to emerge above ground on germination.

Crown Either the basal part of an herbaceous perennial plant from which roots and shoots

grow, or the main branch system of a tree.

Crucifer A plant belonging to the family Cruciferae. All have flowers with four petals arranged to form a cross.

Die-back The death of branches or shoots, beginning at their tips and spreading back towards the trunk or stem.

Dormant Refers to the resting period of a plant, when it makes no new growth.

Exudation Any substance formed in a plant and discharged through a natural opening or wound.

Fasciation Flattening of shoots, sometimes due to the fusion of several stems.

Foliar feed A liquid fertilizer that is sprayed on to, and absorbed through, the leaves.

Foot The base of the main stem of an herbaceous plant.

Fritted trace elements A special formulation of plant nutrients, which will remain available in alkaline soils.

Fruiting body The reproductive body of a fungus, by which it produces spores.

Fungicide A substance used for controlling diseases caused by fungi and some bacteria.

Gall An abnormal outgrowth of plant tissue.

Genus (pl genera) A group of allied species in botanical classification.

Germination The first stage in the development of a plant from a seed, or a fungus from a spore or resting body.

Growing point The site of cell division at the apex of a plant.

Grub see Larva.

Habit The natural mode of growth of a plant.

Heart-wood The central column of dead cells in a tree.

Herbicide Syn for weedkiller.

Host A plant that harbors, or is capable of harboring, a parasite.

Incipient roots Roots that develop from stems; they frequently abort.

Inflorescence The part of a plant that bears the flower or flowers.

Insecticide A substance used to kill injurious insects and some other pests.

Larva The active immature stage of some insects. The larva of a butterfly, moth or sawfly is known as a caterpillar, a beetle or weevil larva as a grub, and a fly larva as a maggot.

Leaching The removal of soluble minerals from the soil by water draining through it.

Lenticel A breathing pore.

Lesion Any localized area of diseased tissue.

Maggot see Larva.

Mosaic A patchy variation of normal green color; usually a symptom of virus disease.

Mottle A variation similar to mosaic, but with small discolored spots.

Mulch A top dressing of organic or inorganic matter, applied to the soil around a plant.

Mutation A change in the genetic make-up of an organism, creating a new, heritable character.

Node The point on a plant stem where a leaf or leaves arise.

Nodule A lumpy outgrowth.

Nymph The active, immature stage of some insects and mites.

Over-winter To pass the winter; it usually refers to the means by which an organism survives winter conditions.

Parasite An organism that lives on, and takes part or all of its food from, a host plant; usually to the detriment of the latter.

Pedicel The stalk of an individual flower.

Pesticide A substance used to kill pests; also, generally, chemicals used to control pests and diseases.

pH The degree of acidity or alkalinity of soil. If the pH of the soil is less than 7, it is acid; above this it is alkaline.

Pollination The transference of pollen from the male to the female parts of a flower.

Proliferation Excess growth usually resulting from abnormally fast cell division.

Pupa In the development of some insects, the resting stage between larva and adult.

Pustule A blister- or pimple-like structure usually produced by a fungus.

Resistant Describes a plant that is able to overcome completely or partially the effect of a parasitic organism or disorder. It also describes a pest or disease that is no longer controllable by a particular chemical.

Resting body A structure produced by a fungus, which remains dormant but viable for a period before germinating.

Rhizome A creeping horizontal underground stem that acts as a storage organ.

Ring pattern Circular areas of chlorosis, the center of each remaining green. It is a symptom of some virus diseases.

Rosette A small cluster of overlapping leaves,

often close to ground level.

Russet Brownish roughened areas on the skin of fruit such as apples.

Scab A roughened, crust-like, diseased area.

Seed dressing A fine powder applied to seeds before sowing to protect them from pests or diseases.

Sepal The outermost, leaf-like structures of a flower.

Snag A short stump of a branch left after incorrect pruning.

Species A group of closely related organisms within a genus. Abbreviations: sp (singular) or spp (plural).

Spore A reproductive body of a fungus.

Spreader A substance added to a spray to assist its even distribution over the target.

Stamen The male reproductive organ of a flower, comprising a stalk with an anther.

Stoma (pl stomata) A minute pore in a leaf surface through which gases and water vapor enter and leave.

Stool The base of a plant, such as a cane fruit, that produces new shoots from ground level.

Strain A distinct group within a species of fungus or eelworm.

Syn Abbreviation for synonym.

Systemic Describes an insecticide or fungicide that is absorbed into the sap-stream and thereby permeates the plant. It also refers to a fungus that is distributed within a plant.

Tilth A fine crumbly surface layer of soil.

Tolerant Describes either a plant that can live despite infection by a parasitic organism, or a fungus that is unaffected by applications of a certain fungicide.

Trace elements Food materials required by plants only in very small amounts.

Transpiration The continual loss of water vapor from leaves and stems.

Tuber A swollen underground stem or root that acts as a storage organ and from which new plants or tubers may develop.

Variety A distinct variant of a species; it may be a cultivated form (a cultivar) or occur naturally.

Watersoaked Describes diseased tissues that look as if they have taken up water or been parboiled.

Wetter A chemical added to a spray to aid wetting of the leaf surface.

Physiological disorders 1

Introduction

Plants are very sensitive to their environment, and there is only a small range of conditions in which a plant will grow successfully. There are five environmental factors upon which plants depend for good growth. These are light, temperature, humidity, water supply and a supply of mineral salts. If any of these conditions are abnormal, then the plant is said to suffer from a physiological disorder. Mechanical and chemical injuries are also discussed in this section, although they are not, strictly speaking, physiological disorders.

Light

Poor light or lack of light causes plants to become thin, weak, drawn and colorless, and they may fail to flower. To a certain extent, such symptoms can arise during the winter, particularly on plants growing in overcrowded greenhouses that do not receive much direct sunlight. If not too severely affected they will recover in good light.

Temperatures

Too high a temperature for storing or forcing bulbs can result in blindness or withering of the flowers. In greenhouses, tomatoes may be affected by greenback, and many plants can show scorching of the leaves. A very hot sun can also cause scalding of fruit. Nothing can be done to prevent such troubles on outdoor plants, but careful ventilation and shading of greenhouses will protect those growing indoors.

Low temperatures at night, even when frost does not occur, can cause young, soft foliage to turn silver, white or yellow, seedlings being particularly susceptible. Cold winds can cause scorching of beech and maple leaves. If frost occurs, leaves may become distorted, especially those of apple, quince and chrysanthemum. The affected leaves curl and the lower leaf surfaces lift so that they can be peeled off easily. Evergreen leaves injured by frost when in the bud or very young become more and more distorted as they grow. Frost can also cause the bark to split and, in severe cases, leaves, shoots and flowers may be killed. Little can be done to prevent frost damage in severe weather but, since weak plants suffer most, encourage vigor by

suitable cultural treatment. Do not grow tender plants in frost pockets. Protect small plants with newspaper or old net curtains when frost is forecast; for half-hardy perennial plants make protective covers by packing bracken between wire netting.

Humidity

Too dry an atmosphere can cause poor growth, bud drop or, occasionally, browning of leaves on house or greenhouse plants. Those that require a humid atmosphere should be syringed daily in hot weather. In centrally heated houses stand pot plants in large containers packed with moist moss or peat. Do not, however, place the pots in saucers of water since this can cause the soil to become waterlogged.

Too moist an atmosphere encourages diseases such as gray mold, potato late blight, tomato blight and tomato leaf mold. Excessive humidity can also cause oedema or dropsy, which shows as raised scab-like corky patches on leaves; it may also affect the shoots and fruits of vines. If oedema occurs, reduce the humidity of the greenhouse by careful ventilation and, when operations such as watering, syringing and spraying have to be carried out, do them in the morning so that the plants can dry out before the evening. Do not remove leaves showing oedema because this will only make matters worse.

Water supply

The water supply for plants must be continuous, but the amount required for any particular stage of growth may vary. Troubles arise when the water supply is deficient, irregular or exceeds a plant's requirements.

A sudden, acute deficiency of water causes plants to wilt, and is likely to occur on a hot day. The following night the leaves' stomata close, transpiration is checked and an affected plant usually recovers its turgidity by morning. No further trouble should occur if the plant is watered. However, if this happens several times, the plant may not recover. Note that wilting can occur rapidly if peat-based soils are allowed to dry out and that such soils are very difficult to moisten again.

Chronic lack of water due to a long drought results in stunting of plants, which bear leaves showing yellow, red or brown tints that are similar to autumn colors. Affected leaves fall prematurely. Root crops and fruit are very small and of poor quality.

A sporadic supply of water causes the development of well characterized troubles such as blossom end rot of tomatoes, hollow heart in potatoes and cracking of other vegetables and fruits.

Lack of water at critical times causes blindness of bulbous plants, dropping of flower buds, flower drop of tomatoes and withering of young cucumber fruits.

Prevent all the above troubles by mulching plants well with farmyard manure, peat, leafmold or other organic materials to conserve moisture. Water in dry periods, even in cold weather if there are cold drying winds, and do not allow the soil to dry out completely. This is particularly important for plants growing against walls since the soil in such positions can remain dry, even in heavy rain if the prevailing wind is not in the right direction. Make sure that pot plants do not dry out, particularly those growing in peat-based soils.

Waterlogging of soil can cause various symptoms such as oedema (see above), or yellowing between the veins, particularly on pot plants and yew trees. Woody plants may develop the condition known as papery bark, in which the bark dries up and peels off. Very occasionally, knobbly gall-like structures appear on the shoots of affected plants; they indicate the development of clusters of adventitious roots. In severe cases discoloration of the foliage and die-back occur. If an affected plant is lifted the roots are seen to be dead and show a purple or blue-black discoloration. The outer tissues may also have peeled off so that only the central cores of the roots are left.

It is not always possible to prevent waterlogging in excess rain, even on light soils. Where waterlogging occurs regularly, however, incorporate some system of drainage if possible, or lift plants and raise the level of the bed by about 6 in before replanting, since this will help to improve the drainage. Try to lighten heavy soils by digging in plenty of

humus and weathered ashes. However, any such treatments must be carried out over a wide area and not just in planting holes. For further information on planting in clay soils see the Introduction, pages 2–3.

Plants affected by waterlogging may be saved if some roots are still alive. Cut off dead and dying shoots and spray the plant with a foliar feed throughout the growing season.

Mineral salts

Many food materials are required by plants and most of them are never likely to be deficient except perhaps in plants that are growing in some soil-less composts and are not fed correctly. Certain nutrients are required in relatively large quantities. These are nitrogen, phosphorus, potassium, magnesium, calcium, sulfur and carbon. They are therefore known as macro- or major nutrients. Other nutrients are just as important but are required in much smaller quantities. They are known as minor or trace elements, or micro-nutrients, and include iron, manganese, boron, molybdenum, zinc and copper.

In gardens, deficiencies of specific major nutrients are less likely to occur than general malnutrition due to neglect. Most trees and shrubs, including those of fruit crops, and also some perennial herbaceous plants, will benefit from annual applications of a complete fertilizer raked in at 3 oz per square yard in spring. Where plants are growing in grass the fertilizer should be applied at double the normal rate—however, a better method of feeding them is to lift small pieces of turf at intervals around the circumference of the plant a foot or so beyond the outermost spread of the branches. Then place a handful of fertilizer in each hole, water this in and replace the turf.

Some types of plants require higher quantities of certain food materials than others, and these may show deficiency symptoms if not fed correctly. For more detailed information on feeding particular crops, see the companion volumes to this book, *Fruit* and *Vegetables*.

When specific nutrients are deficient, certain symptoms are produced, which are listed below. By studying these it is often

Physiological disorders 2

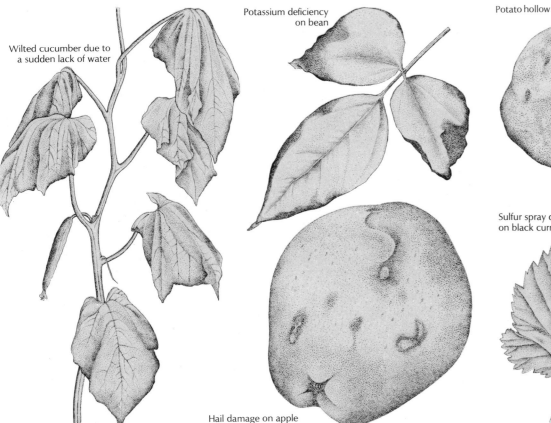

Wilted cucumber due to a sudden lack of water

Potassium deficiency on bean

Hail damage on apple

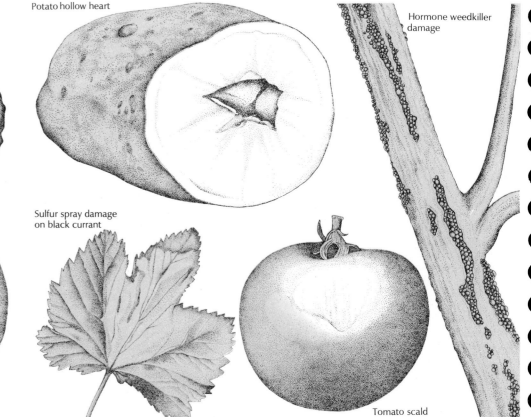

Potato hollow heart

Sulfur spray damage on black currant

Hormone weedkiller damage

Tomato scald

possible to determine the cause of the trouble. However, if in doubt, seek advice and have the soil analyzed.

Nitrogen deficiency symptoms occur mainly on plants growing in light soils lacking organic matter, but they may also develop on pot-bound plants. Both top and root growth are stunted, the shoots are short and thin, and affected plants have a spindly, upright habit. In the early stages of growth the leaves are small and pale yellow but later they develop bright tints of yellow, orange, red and some-times purple. Fruits are small and highly colored, and there is a great reduction in yield of both fruit and vegetables. Correct this trouble by applying nitrogenous fertilizer such as nitrate or ammonium compounds. The rate of application will depend on the type of plant affected. Improve light soils by

digging in a green manure crop such as mustard or rye grass.

Phosphorus deficiency occurs most frequently in those parts of the United States where the rainfall is high, and in areas that have a heavy clay soil. The symptoms are very similar to those caused by a deficiency of nitrogen and are, therefore, not easily differentiated. The main difference is that the leaf color is dull blue-green with purple rather than yellow or red tints, or it may take the form of a dull bronzing with purple or brown spots, as in black currants. Potatoes may develop scorching of the leaf margins. The flesh of fruits from affected plants is soft, acid in flavor and does not keep well, and the skin is green but may be highly flushed.

Prevent these troubles on fruit by applying

superphosphate at about 1½ oz per square yard every second or third year. Bone meal is high in phosphorus and a dressing of this at 4 oz per square yard prior to sowing or planting will benefit beds of annuals and bearded iris.

Potassium deficiency symptoms may occur on plants growing in clay soils if the plant requires a lot of potash, for instance, potatoes, tomatoes, beans and fruit crops. More frequently, however, this deficiency occurs on light sandy, peat or chalk soils. Affected plants are stunted and bear dull blue-green leaves which may show browning, either as small spots or at the tips or around the margins. The leaves may also curl downwards. For those fruit crops requiring a lot of potassium, such as raspberries, apply sulfate of potash annually in March at

¾ oz per square yard. For poor flowering or berrying trees and shrubs apply this fertilizer in late summer or early spring at about ½ oz per square yard. Note that tomato fertilizers are usually high in potash.

Magnesium deficiency symptoms occur very frequently on all types of plants because this element is easily leached from soils during heavy falls of rain. It is also made unavailable by excess potash in the soil, and deficiency symptoms often appear on tomatoes and chrysanthemums that are fed with high potash fertilizers or where a lot of bonfire ash is raked in, since wood ashes are high in potash. The symptoms appear first on the older leaves and spread progressively up-wards. Chlorosis (yellowing) is the commonest symptom, but brilliant orange-brown and red tints may also develop, and the leaves

Physiological disorders 3

fall prematurely. Apply magnesium sulfate to the soil at 1 oz per square yard. Alternatively, for a quicker response spray the foliage with a solution of $\frac{1}{2}$ lb magnesium sulfate in 3 gal of water to which is added a spreader-sticker such as soft soap or a few drops of mild liquid detergent. For most plants spray two or three times at two week intervals. Other spraying times are given under specific hosts.

Manganese deficiency is prevalent in some areas of the United States. Soils that are generally deficient in manganese are sands, alluvial silts and clays. A deficiency of manganese can also be induced where the pH is over 7.5 and, therefore, the trouble frequently occurs together with iron deficiency (see below). Deficiency symptoms may appear following heavy rain since, in some circumstances, manganese is made unavailable if the soil is very wet. In general the symptoms are very like those caused by a deficiency of magnesium, that is, chlorosis of the leaves, but more specific symptoms are shown by peas and beets, for which see the relevant pages. Where manganese deficiency is suspected, spray with a solution of manganese sulfate at the rate of 2 oz in 3 gal of water plus a spreader, repeating two or three times at two week intervals. Alternatively apply a chelated compound or fritted trace elements to the soil.

Iron deficiency is almost always induced by very alkaline soil conditions in which the pH is over 7.5. For this reason the condition is often known as lime-induced chlorosis. Young growths are always the most severely affected and scorching of the leaf margins and tips occurs in extreme cases. Sometimes the symptoms are so severe that the leaves are white. In milder cases, however, it may be difficult to distinguish between iron, manganese and magnesium deficiencies; in this case it is essential to determine the pH of the soil. If the soil is acid the symptoms are nearly always due to a deficiency of magnesium (see above). Lime-induced chlorosis is likely to occur on acid-loving plants such as rhododendrons, camellias and most heathers where the soil pH is greater than 7.0. If it is above 8.0, ceanothus, chaenomeles, hydrangeas, raspberries and peaches are also likely to

show symptoms. Try to reduce the pH of the soil by digging in acidic materials such as peat, pulverized bark or crushed bracken, or use acidifying chemicals. Flowers of sulfur can be used on sandy loams at 4 oz per square yard. Try a small quantity first, test the pH after a few months and repeat applications as necessary until the pH is reduced to the required level. Aluminum sulfate and ferrous sulfate can also be used, but only where the ground is vacant, and it should not be planted up until the soil has reached the required pH. Apply either chemical at 4–8 oz per square yard, rake in well and water in dry weather. After two or three weeks test the soil pH and repeat if necessary. Use of aluminum or ferrous sulfate could result in phosphate being made unavailable (for symptoms of this see under Phosphorus deficiency). Even with these treatments, it is unlikely that it will be possible to reduce a pH of over 7.5 to as low as 5.5, which some acid-loving plants require, so apply annually a proprietary product containing chelated compounds, or use fritted trace elements.

Calcium deficiency does not usually affect plants in gardens, even on acid soils. In very acid, peat-based soils in growing bags, however, it may cause tomatoes to suffer from blossom end rot. Calcium deficiency within the fruit can cause bitter pit of apples. For details, see page 15.

Boron deficiency may be due to lack of boron in the soil or it can be made unavailable by over-liming. It produces specific symptoms in beet, swedes, turnips and pears. For control measures see under the appropriate host plant.

Molybdenum deficiency occurs occasionally in acid soils. Symptoms appear only on brassicas, where the deficiency is known as whiptail. For control measures, see the section on brassicas, page 34.

Combination of adverse factors

Many of the troubles listed above produce similar symptoms, and it is sometimes difficult to determine the exact cause of the trouble affecting a plant. Furthermore, one unsuitable cultural condition can result in other troubles so that a plant may suffer from a combination of adverse factors. Thus, roots injured or even

killed by drought are unable to take in food materials, and so the plant will not only be affected by the lack of water but by malnutrition as well. Similarly, waterlogging can cause the roots to die and also results in poor aeration of the soil, so that the roots are unable to take in oxygen. Poor planting will result in inefficiency of the root system and the plant may be unable to absorb sufficient water and food materials. Throughout this book, therefore, wherever faulty root action is mentioned it indicates either one of the troubles above or any combination of them.

Occasionally, specific disorders can arise as a result of a combination of adverse factors; these are mentioned under the specific hosts. For example, bitter pit of apples and blossom end rot of tomatoes are due to a deficiency of calcium within the fruits, although it usually arises as a result of irregular watering. Similarly, shanking of grapes is due to overcropping of a vine whose root system is functioning inadequately.

Mechanical injury

Man-made injuries to plants are usually obvious, for example wounds caused when trees are hit accidentally by motor mowers or when plants are damaged by hoeing. Any large wound on a woody plant should be cleaned up so that no ragged tissues are left. Potentially more serious are small wounds that are easily overlooked, for example pruning snags and tight ties around stems. The latter can result in die-back of shoots on trees and shrubs. Keep a careful watch on ties and labels and, as the plants grow, loosen them so that they do not become embedded in the tissues and strangle the shoots.

Natural injuries are caused by extreme weather conditions. Heavy snow can break branches, so remove snow from conifers as frequently as possible. Nothing can be done, however, to prevent wind damage, which can also break branches and tatter large leaves such as those of horse chestnut. Hail can cause holes in leaves or small pits in fruit and canker-like injuries on woody stems.

Chemical injury

This type of damage occurs when an insecticide or fungicide is applied to a plant that is

sensitive to the chemical. For example, sulfur sprays can cause russeting of fruit on some apple varieties, defoliation of some types of gooseberries and brown blotches on black currant leaves. Copper sprays can cause scorching of the leaves on weak rose bushes and on rhododendrons. Plants that are sensitive to a particular chemical are usually listed on the container. Therefore, always read the label before applying a chemical to ensure that it can be used safely on that type of plant.

Most chemical injuries are caused either by drifting, which may occur if weedkillers are used on a windy day, or by applying pesticides with apparatus that was badly rinsed after being used to apply a weedkiller. Keep special equipment for applying weedkillers and do not wash it out in a water butt or tank used for irrigating plants. Most damage to plants is caused by hormone weedkillers such as 2,4-D and dicamba, and symptoms can arise on plants that have been mulched with compost containing cuttings from a lawn treated with one of these weedkillers. Do not put the first mowings after treatment on a compost heap, nor use them as a mulch. The symptoms caused by hormone weedkillers are twisting and distortion of the shoots and leaves, which may show cupping. In severe cases, however, small nodules may appear on the shoots of shrubs or herbaceous plants such as chrysanthemums due to the development of adventitious roots. Plants affected by those hormone weedkillers that are available to amateurs usually grow out of the symptoms in due course.

Non-hormone weedkillers can also damage plants. For example, aminotriazole and simazine cause either chlorotic blotches or yellowing between the leaf veins. Paraquat/diquat can cause complete yellowing of the leaves of bulbous plants the following season if the weedkiller is applied before the foliage has died down completely and while the necks of the bulbs are still open. Prevent this trouble by covering the bulbs with soil before treating the area. Use all weedkillers with care and follow the manufacturer's instructions concerning their use, particularly with regard to the types of plants that may be injured.

Tree fruit: leaves 1

This section covers apples, pears, plums and related fruits, cherries, peaches, nectarines, apricots and quinces grown for their fruit. For ornamental trees, see pages 83–90.

Introduction

The troubles dealt with in this section are those that reduce yields, damage the fruits or interfere with the growth of young plants. Fruit crops are also susceptible to a number of physiological disorders that are not mentioned specifically here. It is advisable, therefore, to read the section on physiological disorders (pages 5–7) when trying to identify the cause of the trouble that is affecting any unhealthy fruit crop.

Few physiological disorders should arise if fruit crops are given the correct cultural treatment needed to produce good crops. However, health cannot be achieved unless good quality fruit trees, bushes and canes are planted; whenever possible buy plants that are certified to be healthy, in particular that they are free of viruses, and true to type. Since most fruits are grown as perennials their permanence on a particular site favors a build-up of pests and diseases. Prevent such a build-up by good hygiene, and observe the basic principles listed in the Introduction on pages 2–3.

Leaves with pests visible

Aphids (various species) attack all types of fruit, especially in the late spring and early summer. Most species over-winter on the tree as eggs, which hatch in the spring as the buds start to open. The aphids may have varying colors, notably pale green, gray-pink or black. Heavy infestations develop at the shoot tips, sometimes causing leaf curling and stunted growth. The sugary excretions produced by aphids make the foliage and fruit sticky and can allow the growth of a black sooty mold. Aphid infestations normally die out in late June or July when winged aphids develop and leave the fruit tree to migrate to their summer hosts, which are often wild plants. There is, therefore, little point in spraying in midsummer when symptoms are most obvious, since by then the damage will already have been done. Spray the trees in December or January with a dormant oil wash to control over-wintering eggs. In the spring, apply a contact insecticide to apple, cherry, plum and quince when the flower buds appear, but before they open, while pear and peach should be sprayed when flowering has finished to catch any recently hatched aphids which have escaped the winter treatment.

Pear and cherry slugworm caterpillars (*Caliroa cerasi*) grow up to $\frac{5}{8}$ in long and resemble slugs since they are black and slimy. They feed by grazing away the surface tissues of leaves, mainly on the upper surface but sometimes on the undersides. The leaves are not holed but the damaged areas dry up and turn brown. There are two or sometimes three generations during the summer, and caterpillars can be found on the foliage of pears and cherries between late June and October. Light infestations cause little harm but if many leaves are affected the trees can be sprayed with derris or carbaryl.

Leaves with holes

Shothole is caused by the fungus *Coryneum carpophilum* and attacks only cherry, peach, nectarine, plum and gage trees that are lacking in vigor. Numerous brown spots develop on the leaves and the dead tissues later fall away leaving holes. Prevent this disease by feeding, mulching and watering trees to increase their vigor. Should symptoms appear spray with a foliar feed during the growing season. If the disease recurs later in the season, spray affected trees with a solution of captan or dodine.

Larvae of the winter moth (*Operophtera brumata*), spring cankerworm (*Paleacrita vernata*) and fall cankerworm (*Spodoptera frugiperda*) feed on the newly emerged leaves, blossoms and fruitlets of all tree fruits during the spring. These caterpillars walk with a looping action and bind leaves loosely together with silken threads. The small holes that these caterpillars make in the young leaves often go unnoticed at the time of feeding but are easily seen when the foliage is fully expanded. By then the caterpillars have finished feeding and have gone down into the soil to pupate. The adult insects emerge from late October to early April and are unusual because only the males are capable of flight. Females merely have stumps for wings and have to crawl up the tree trunk or stake to reach the branches and lay eggs. Reduce the number of females that lay eggs by placing a sticky grease band around the trunk and stake, keeping it sticky and free of dead leaves throughout the emergence period. The band should be 3–5 ft above ground level and 5–6 in wide. Winter washes give some control of the eggs but it is better to spray against the young larvae by applying fenitrothion or trichlorphon at the green cluster stage on apple, at white bud on cherry, quince and plum, and at petal fall on pear and peach.

Leaves distorted

Aphids (various species) frequently cause leaves to become curled and crinkled. These symptoms occur on the leaves at the shoot tips, usually between May and early July. If damaged leaves are uncurled during the early part of this period the aphids can usually be seen, but later there may be nothing left except cast skins, since by then the aphids will have migrated on to the herbaceous plants on which they feed during the summer. Their control is described above under Leaves with pests visible.

Peach leaf curl (*Taphrina deformans*) affects only peaches, nectarines, almonds and

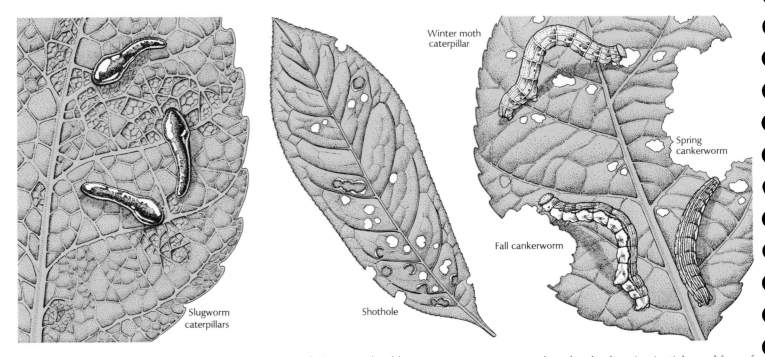

Slugworm caterpillars

Winter moth caterpillar

Spring cankerworm

Fall cankerworm

Shothole

Tree fruit: leaves 2

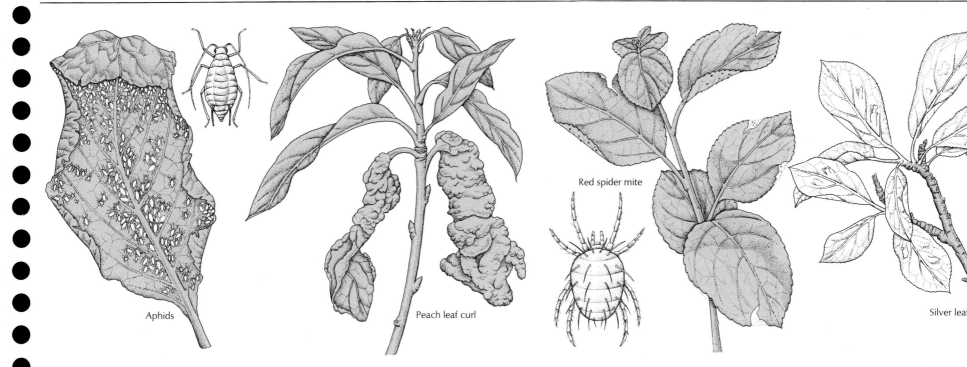

Aphids

Peach leaf curl

Red spider mite

Silver leaf

occasionally apricots. All or part of the leaf becomes considerably thickened and curled, and blisters develop. The blisters are red at first and later become covered with a pale bloom consisting of masses of fungus spores. On small trees, or where only a few leaves are swollen, reduce the infection by removing these leaves before they turn white. Where a number of leaves are infected they fall very early in the season and weaken the tree. The fungus dies out in the fallen leaves, but over-winters as spores lodged in cracks and crevices in the bark, on the shoots or among the bud scales. Prevent germinating spores from entering the buds by spraying with ferbam, zineb or a copper fungicide in January or February. Repeat 10–14 days later and also just before leaf-fall.

Leaves discolored
Fruit tree red spider mite (*Bryobia rubrioculus*) attacks the foliage of apple and plum, while peach is susceptible to greenhouse red spider mite (*Tetranychus urticae*). These mites are only just visible to the naked eye and a hand lens is necessary to see them clearly. They

live on the undersides of the leaves where they feed by sucking sap. Heavy infestations can occur in hot dry summers. The foliage becomes dull green at first but later may become mottled or turn yellow, and fall prematurely. *B. rubrioculus* over-winters on apples and plums as red spherical eggs which are laid on the underside of smaller branches. Sometimes these eggs are sufficiently numer-ous to give the bark a dark pink color. Control these eggs by applying a wash containing DNOC in dormant oil in early February. During the summer control these mites with malathion or systemic insecticides, with a first application in late May and again three to four weeks later. If mildew also needs to be controlled on apple, a good fungicide is dinocap as this also helps to suppress the mites. *T. urticae* over-winters as adults under flakes of bark and is not so susceptible to winter treatments. Check it by using the summer insecticides described above.
Silver leaf is caused by the fungus *Stereum purpureum* and enters through wounds such as pruning snags. It can affect any type of fruit, though plums are most susceptible to

infection, especially the variety 'Victoria'. The fungus passes upwards in the sap and causes the leaves on one or more branches to be-come silvered and then sometimes brown. Progressive die-back of affected branches occurs and small fruiting bodies of the fungus develop on the dead wood. These are purple when fresh, later becoming white or brown. They may be bracket-shaped, overlapping or they may lie flat. A brown or purple stain develops in the internal tissues and is easily seen if the cut end of an affected branch at least an inch in diameter is moistened. Cut back all dead branches to about 6 in beyond where the stain ceases. It is preferable to do this during summer when there is the least chance of new infections occurring. All pruning tools should be sterilized both before and after use, and when the work is com-pleted paint the wound with a fungicidal paint. If fruiting bodies of the fungus appear on the trunk of the tree, then the tree will have to be destroyed. If only the branches are affected, then the tree should recover if it is fed, mulched and watered to encourage vigor.

False silver leaf is a physiological disorder that is caused by malnutrition or an irregular supply of moisture in the soil. It is probably more common than true silver leaf and is often confused with it. The foliage again shows a silvery discoloration but, unlike silver leaf, most leaves on the tree are affected at the same time and there is little or no die-back. If such symptoms appear early in the season spray with a foliar feed. The following year feed, mulch and water as necessary to encourage vigor. Should any die-back occur cut out branches back to the clean living tissues and paint the wounds with a fungi-cidal paint.
Honey fungus (*Armillaria mellea*) causes the leaves to become discolored and the shoots to die back. Affected trees often die suddenly. For a description of the symptoms and treat-ment of this disease, see under Roots dead, page 17.

Leaves spotted
Pear leaf blister mite (*Phytoptus pyri*) starts to damage the leaves of pear, and occasionally apple, in May. The young leaves develop pale

Tree fruit: leaves 3

green or pink blotches which are often arranged in rows along either side of the mid-rib. As the summer progresses the blotches turn brown or black. The mites live inside the discolored areas of the leaf and are too small to be seen without a microscope. The leaves may become heavily disfigured but no real harm is caused to the fruit, nor to the tree's vigor. There are no effective chemicals available to amateur gardeners at the present time that will control this mite; light infestations can be checked by removing affected leaves or small branches but this is not worth while if pear leaf blister mite is present throughout the tree.

Apple scab and pear scab (*Venturia inaequalis* and *V. pyrina*) show as olive-green or brown blotches on leaves, which fall prematurely. Rake these up and burn them to prevent overwintering of the fungi. Spray with benomyl, captan or dodine at bud burst and repeat every two weeks as necessary until late July, or use captan at the green cluster stage. For apples repeat treatment with captan at the pink bud stage, then when 80 per cent of the petals have fallen and again three weeks after petal fall. For pears repeat at the white bud stage, at petal fall and at the fruitlet stage. Then repeat every two weeks.

Quince leaf blight (*Entomosporium maculatum* syn *Fabraea maculata*) produces small irregular spots which are red at first but later turn almost black. The spots may coalesce and the leaves turn brown or yellow and fall prematurely. Spotting and distortion of the fruit may also occur and infected shoot tips die back, but these symptoms are not as common as the leaf spots. Rake up and burn fallen leaves and cut out dead shoots in winter. Spray diseased trees when the first leaves appear in spring using benomyl or a copper fungicide. Repeat during the summer if further symptoms appear.

Leaves with visible fungal growth

Powdery mildew of apples, quince and pears (on which it is rare) is due to *Podosphaera leucotricha*. On peach it is caused by *Sphaerotheca pannosa* f *persicae*. In both cases it shows as a white powdery coating of fungus spores on the emerging leaves in spring, causing them to wither and fall. The disease

PHYSIOLOGICAL DISORDERS

Physiological disorders may cause discoloration of leaves different from that produced by false silver leaf. All such discolorations are due to unsuitable soil conditions. Cherry and peach leaves frequently show yellowing or even whitening between the veins due to the soil being too alkaline—a trouble known as lime-induced chlorosis. Apples are particularly susceptible to magnesium deficiency which produces orange-yellow blotches between the veins. It is suspected that magnesium deficiency may also be partly responsible for Cox's spot which shows as small, round, pale brown spots on the leaves. This trouble is worse on trees suffering from faulty root action as a result of drought or waterlogging. Adverse soil conditions can also cause black blotches on pear leaves. Whenever leaves show discoloration consult the section on physiological disorders (pages 5–7).

spreads to later developing leaves. Cut off infected shoots in spring and summer. Spray apples and quinces with dinocap at the pink bud stage and repeat every two weeks until mid-July. Alternatively use benomyl or thiophanate-methyl at the green cluster stage and repeat every two weeks. On peaches use a sulfur fungicide at the first signs of trouble and repeat treatment every two weeks as necessary.

Plum rust (*Tranzschelia pruni-spinosae* var *discolor*) can also attack apricots, peaches and nectarines as well as plums but is of little importance except in hot dry seasons when large numbers of leaves may fall prematurely. The first symptom is the occurrence of yellow spores on the undersides of leaves. The disease then spreads to other leaves via the spores. Later, brown over-wintering spores develop. Therefore rake up and burn fallen leaves to reduce the infection. Diseased plums may be sprayed with thiram or zineb but it is much more important to reduce susceptibility of the trees by feeding, mulching and watering them. In dry periods ensure that the soil does not dry out completely.

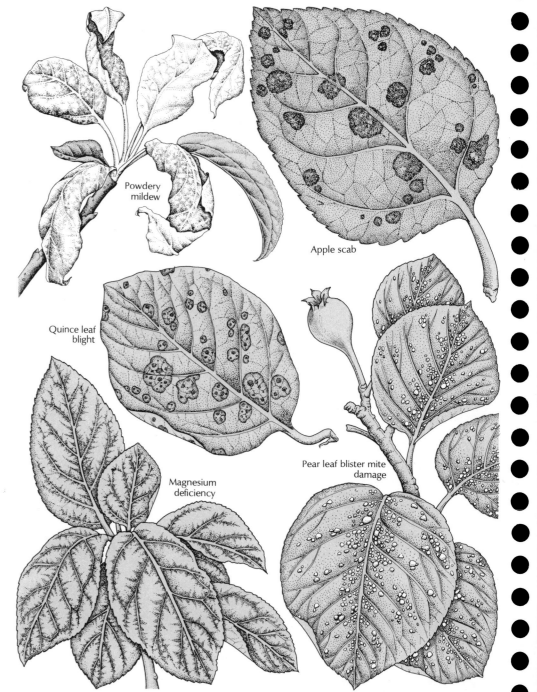

Powdery mildew

Apple scab

Quince leaf blight

Magnesium deficiency

Pear leaf blister mite damage

Tree fruit: flowers

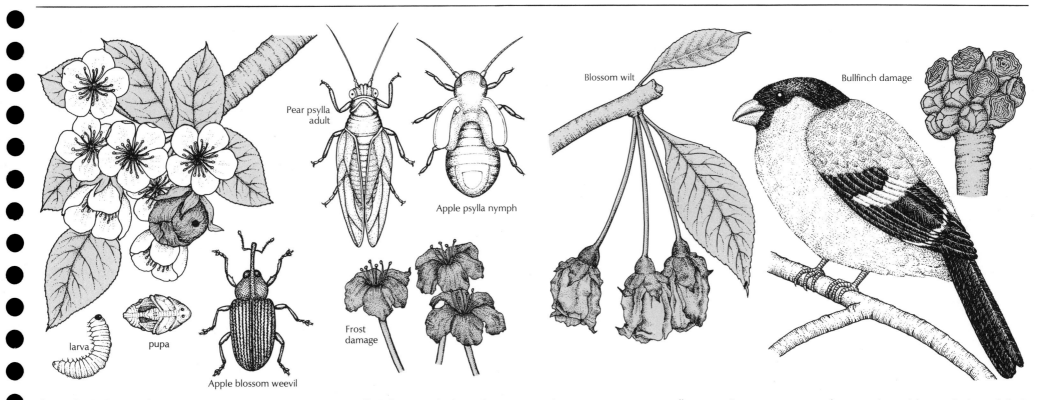

Pear psylla adult

Apple psylla nymph

larva

pupa

Apple blossom weevil

Frost damage

Blossom wilt

Bullfinch damage

Flower buds damaged

Bullfinches (*Pyrrhula pyrrhula*) eat the unopened flower buds of cherries, plums and pears during the winter, and attack apple blossom between bud burst and the pink bud stage. In some years attacks are so severe that only a few blossoms escape. Bullfinch damage can be detected by the discarded bud scales, which are easily seen below the tree when the ground is covered with frost or snow. An examination of the twigs will show that many of the buds have been nipped off, although those at the shoot tip usually remain. This pattern of bud removal can be seen in the spring when branches are bare of blossom except at the shoot tip. Different varieties of fruit vary in their susceptibility but none is completely immune. The pear variety 'Doyenne du Comice' often escapes damage. Protect other varieties with bird repellant sprays or some form of netting in areas where bullfinches are a problem. Netting or a fruit cage is the most effective method where the trees

are not too tall. Otherwise the branches can be draped with rayon fibers such as Scaraweb. Various bird repellant sprays are available but these have to be applied on several occasions, especially in wet years, and are only effective while there is an alternative source of food in the locality.

Flowers damaged

Apple psylla and pear psylla (*Psylla mali* and *P. pyricola*) suck sap from the blossom trusses and young leaves of apple and pear respectively. Heavy infestations can kill the blossom, and the brown color of the petals may be mistaken for frost damage. Psyllas are pale green or yellow-brown insects and their nymphs are about $\frac{1}{10}$ in long and flattened with large wing pads. Apple psylla has only one generation a year and the adults, which suck sap from the foliage later in the summer, cause little damage. Pear psylla, however, has at least two generations and can soil the foliage and fruit with honeydew and sooty mold. Unlike apple psylla, which over-

winters as eggs, pear psylla over-winters as adult insects. Control apple psylla eggs with dormant oil spray in the winter, or kill the newly hatched nymphs with a systemic insecticide applied at the green cluster stage of bud development. Infestations of pear psylla may need spraying with a systemic insecticide at petal fall but, if required, spraying can usually be delayed until three weeks later.

Apple blossom weevil (*Anthonomus pomorum*) is a less common problem now than in former years. The small brown beetles lay their eggs in the unopened flower buds. These hatch into small white grubs that eat the central parts of the flower. Infested blossoms fail to open and, after petal fall, the capped blossoms with their brown petals are readily seen. When examined these flowers will either contain the beetle's grub or pupa, or there will be a hole in one of the petals where the newly formed adult has emerged. In gardens where blossom weevils have been troublesome in previous years, control these

pests by spraying with a solution of fenitrothion at the bud burst stage.

Flowers withering

Frost damage may cause petals to turn brown and the flowers to drop if they are fully open at the time. Flowers injured while in bud may not open fully and just show one or two withered petals. No treatment can be carried out to prevent this trouble apart from avoiding planting in frost pockets or using varieties that are either late-flowering or to some degree tolerant of frost.

Blossom wilt (*Monilinia* spp) can affect all types of tree fruit. It is a seasonal disease and is most troublesome in wet springs. All the flower tissues on an infected shoot wither, but they do not fall. The leaves may also shrivel and the tip of the shoot die back. Cut out and burn all infected shoots in summer. The following season help to prevent the disease from recurring by spraying all fruit trees with benomyl as the first flowers open. Repeat treatment seven days later.

Tree fruit: stems 1

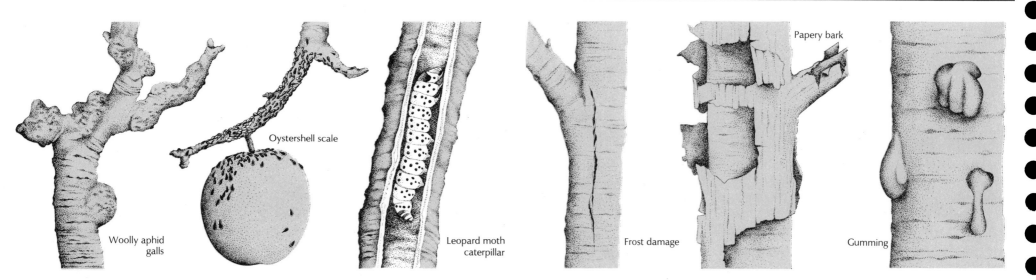

Woolly aphid galls

Oystershell scale

Leopard moth caterpillar

Frost damage

Papery bark

Gumming

Stems with pests visible

Woolly aphid (*Eriosoma lanigerum*) appears during the spring on the bark of apple branches, especially around old pruning cuts or where the bark is split. The aphids are gray-brown and covered by white fluffy wax fibers. These can be mistaken for a mold but if a finger is rubbed over the colony it becomes wet with crushed aphids. Later in the summer woolly aphids spread to the new shoots, especially soft growths arising from the trunk. Where the aphids have been feeding, the bark develops soft lumpy galls which are most readily noticed in the winter. These galls are liable to be split open by frost and allow the entry of apple canker spores. Control woolly aphid by applying a malathion insecticide as soon as the pest is seen. This can be sprayed on the tree, but light infestations can be treated more effectively by applying the insecticide to the aphid colonies with a paint brush.

Scale insects such as oystershell scale (*Lepidosaphes ulmi*) attack apple, while peach scale (*Lecanium persicae*) occurs on peach. The pests suck sap and live under protective shell-like coverings which encrust the bark. Oystershell scale is about $\frac{1}{8}$ in long and resembles the shellfish of the same name. Peach scale has a convex, oval shell which is up to $\frac{1}{4}$ in long. Control them by spraying with dormant oil. Alternatively, spray with mala-

thion in early June against oystershell scale or early July for peach scale.

Stems with peeling bark

Papery bark is seen most commonly on apples and is due to faulty root action, in particular waterlogging. The bark becomes paper thin and peels off as a pale brown sheet. Many shoots may be affected and, if they are girdled, die-back frequently occurs. Cut out all dead shoots and on the larger branches remove dead and rotting tissues beneath peeling bark. Then cover all wounds with a fungicidal tree paint. Improve the cultural conditions to prevent further trouble.

Stems tunneled

Leopard moth (*Zeuzera pyrina*) is a fairly widespread insect, but is not a major pest since it is rarely found in large numbers. Apple is the most susceptible but all fruit trees can be attacked. The outward sign of an attack is a hole in a branch or trunk, which is the entrance to a tunnel that may penetrate the wood for more than 6 in. Pellets of sawdust frass often accumulate around this hole. Inside the tunnel will be found a caterpillar with a brown head and a creamy-yellow body with black spots. It grows to nearly 2 in long and attacks branches and trunks up to 4 in in diameter. The tunneling can cause small branches to die, but the

damage may not be apparent until strong winds cause the weakened branch or trunk to break. If signs of tunneling are seen the caterpillar can be killed either by pruning out the affected branch or by pushing a piece of wire into the tunnel.

Stems with split bark

Frost damage and irregular growth due to faulty root action can both cause longitudinal splitting of the bark. Affected shoots may show discolored foliage or may even die back. Clean the wound by cutting out any rotting tissues, and cover the tissues with a wound paint. Also, remove any dead wood and feed, mulch and water or drain the soil as necessary.

Stems gumming

Gumming from otherwise healthy shoots of plum, peach or cherry trees indicates lack of vigor caused by malnutrition or unsuitable soil conditions. If, however, the gumming occurs on a stem which appears flattened, it is a symptom of bacterial canker (*Pseudomonas syringae*). The gum is usually pale or dark brown and viscous at first, later setting into hard, irregularly shaped lumps. It very often dries up on its own but if further exudation occurs the gum should be removed leaving a clean wound, which should be covered with wound paint. Encourage vigor

in the affected tree by feeding, mulching and watering as necessary. If the trouble is due to bacterial canker, see below. In this case some die-back is likely to occur.

Stems cankered

Bacterial canker (*Pseudomonas syringae* or *Pseudomonas mors-prunorum*) is the most serious disease of plums and can also be troublesome on peaches and cherries. The cankers show on branches as elongated flattened lesions from which exude copious amounts of gum. The following spring the buds of an affected branch fail to open, or if leaves do develop they turn yellow and become narrow and curled, then wither and die during the summer as the branch dies back. Another symptom of the disease is brown circular spots on the leaves with the affected tissues falling away to leave holes resembling those of shothole (see page 8), but this is less serious than the die-back of the cankered branches. Very occasionally the bacteria die out in the cankers and no further symptoms are seen. However, badly cankered branches and dead wood should be removed and the wounds painted with a fungicidal wound paint. The bacteria live on the leaves during the summer, therefore spray the foliage thoroughly with bordeaux mixture in mid-August, mid-September and mid-October.

Tree fruit: stems 2

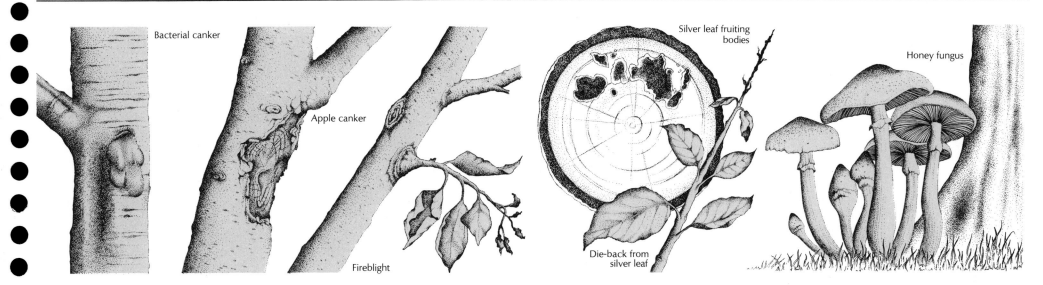

Bacterial canker

Apple canker

Fireblight

Silver leaf fruiting bodies

Die-back from silver leaf

Honey fungus

Apple canker (*Nectria galligena*) is a very destructive disease of apples, and can also affect pears. Sunken and discolored patches develop on the bark and as they extend become elliptical with the bark shrinking in concentric rings around each canker. The branch usually becomes swollen around the canker and girdling of the shoot may occur, which causes die-back. In summer white pustules of fungus spores form on the sunken bark, and later small red fruiting bodies develop by which the fungus over-winters. Infection can occur at any time of the year through wounds, pruning cuts and leaf scars. Cut out severely cankered branches and spurs. On large branches where no die-back has occurred cut out the brown diseased tissues with a chisel or sharp knife and collect the prunings. Paint the wounds with a benomyl tree dressing. If severe apple canker occurs on smaller trees, spray with bordeaux mixture or liquid copper after harvest but before leaf-fall, then at 50 per cent leaf-fall and again in spring as the buds burst. Since the disease is worse on trees lacking in vigor feed and mulch them and improve the drainage if any waterlogging occurs. This is particularly important for varieties susceptible to canker such as 'Cox's Orange Pippin', 'Lord Suffield' and 'James Grieve'.

Fireblight (*Erwinia amylovora*) affects apples and pears. In the fall, cankers can be seen at the base of dead shoots, and if the bark is pared off a red-brown discoloration can be seen in the inner tissues. The bacteria usually enter through late spring or summer blossoms in warm weather. They spread down the spurs and lateral twigs into the main branches causing die-back and browning of the leaves, which become withered but do not fall. In spring the cankers become active again and droplets containing bacteria ooze out of them and are carried by rain and insects to the flowers as they open. If this disease is noticed, spray with streptomycin when the first flowers are at the ballon stage (ie when they are shaped like a brandy glass) and repeat when fully open. Spray a third time during extended flowering periods. Inspect trees in late summer and cut out cankered branches to a point well below the diseased area. Disinfect all pruning tools after use. Continue inspecting the trees during the fall and winter, cutting out any cankered areas that were previously missed. A final inspection in March is also desirable.

Stems with small eruptions
Tan bark is a physiological disorder that shows on one-year-old shoots of cherry trees, or sometimes the main trunk, as small eruptions containing a tan-colored powder. These eruptions are the breathing pores (lenticels) of the tree and are prominent

naturally, but on affected trees they burst open causing the outer layers of the stem to peel off and exposing brown rust-like masses of dead plant cells. Although unsightly, this trouble is not serious since it usually occurs where the root action is vigorous, although it can also be due to waterlogging of the soil. Apart from draining the soil no treatment is required unless much of the bark has peeled. In this case remove the loose bark and also any dead tissues beneath to leave a clean wound. Cover this with a fungicidal paint.

Stems dying back
Die-back can be due to silver leaf, apple canker, bacterial canker, fireblight and honey fungus. Determine the cause by examining trees for other symptoms of these diseases. If no definite symptoms of disease can be seen, the die-back is probably due to unsuitable soil conditions and appropriate remedial action should be taken.

Honey fungus (*Armillaria mellea*) frequently causes fruit trees to die suddenly. The primary symptom is die-back of the stems, the leaves of which wither but continue hanging for a while before falling. The roots are also affected. For their symptoms, see under Roots dead. In the fall, honey-colored toadstools appear at the base of the dying tree. Dig up and burn dead and dying trees together with as many of the roots as possible. Sterilize the

soil with a 2 per cent solution of formalin applied at the rate of 6 gal per square yard, or with a proprietary phenolic emulsion.

Stems with visible fungal growth
Powdery mildew of apple (*Podosphaera leucotricha*) and of peach (*Sphaerotheca pannosa* f *persicae*) shows on both shoots and leaves as a white mealy coating of fungus spores. For symptoms and treatment of powdery mildew on leaves, see page 10. Infected shoots are usually stunted and may die back at the tips. Shoots may still show signs of fungal growth in the winter and they should be cut out to a point two or three buds behind the diseased tissues. Control powdery mildew on apples and quinces by spraying with dinocap at the pink bud stage, repeating every two weeks until mid-July. Alternatively apply benomyl or thiophanate-methyl at the green cluster stage, repeating every two weeks. On peaches use a sulfur fungicide at the first signs of trouble, repeating every two weeks as necessary.

Woolly aphid (*Eriosoma lanigerum*) is a gray-brown aphid that is covered with white fluffy wax fibers, which give colonies the appearance of a fungus. Woolly aphid is distinguished by rubbing the "fungus" with a finger; if it is woolly aphid the finger becomes wet with crushed insects. For treatment see under Stems with pests visible.

Tree fruit: fruit 1

Fruits lacking

Frost damage and blossom wilt may, because of their effect on bud and flower (see page 11), prevent fruit from appearing. Another common cause is lack of pollination of the flowers. This may be due to adverse weather conditions discouraging the pollinating insects, or because there is no tree of a suitable cross-pollinating variety in the vicinity. Do not plant just one fruit tree without first checking that it will produce fruit in the absence of cross-pollination.

Fruits dropping

June drop of apples is a natural occurrence and should cause no concern. When fruitlets of any type drop early the trouble is usually due to lack of effective pollination (see Fruits lacking). Premature dropping of larger fruits can be due to pest damage, therefore examine the fallen fruits for maggots. If the fruit is undamaged, then the cause is probably faulty root action. Take appropriate remedial action the following season.

Fruits eaten

Birds and wasps feed on the ripe fruit of most tree fruits. On tough-skinned fruits such as apples, wasps usually enlarge damage started by birds, but they can initiate damage on the softer-skinned fruits such as plums and pears. Deal with wasps by searching for their nests and destroying them by tipping some carbaryl dust into the entrance at dusk. Keep birds away from small trees by growing them inside a fruit cage. On larger trees enclose individual fruit trusses in bags made of muslin or old nylon tights. These have the advantage of helping to keep wasps away.

Fruits distorted

Frost damage when fruitlets are developing sometimes causes them to distort. Apple fruits may become pear-shaped, or narrower around the middle than at either end. No action can be taken to prevent frost damage except to avoid planting fruit trees in frost pockets. If this must be done plant either late-flowering varieties or those that are reasonably tolerant of frost.

Stony pit is a virus disease of pears. The fruits are pitted and deformed at maturity and have patches of dead stony cells within the flesh, making the fruits inedible. The trouble may show first on one branch only but will gradually spread to all branches, particularly if the tree is old. Destroy infected trees, and buy only trees certified to be free of viruses.

Boron deficiency causes distortion of pears on most branches of a tree. Affected fruits have brown spots in the flesh and the bark has a roughened, pimpled appearance. These symptoms are usually accompanied by dieback of some shoots, while on the others the leaves are small and misshapen. Where this is a recurrent problem spray at petal fall with 2 oz of borax (sodium tetraborate) in 5 gal of water to which is added a spreader.

Fruits cracked

Irregular supply of moisture at the roots can cause cracking in fruits that are otherwise healthy. It can be prevented to a certain extent by mulching and by watering in dry periods before the soil dries out completely.

Fruits with blemishes

Apple scab and pear scab (*Venturia inaequalis* and *V. pyrina*) produce brown or black scabs on the fruit. The scabs may be numerous and almost cover the fruit. In such severe cases the surface becomes corky and may crack. Spray infected trees with benomyl or dodine at bud burst, repeating treatment every two weeks until late July. Alternatively apply captan at the green cluster stage. Then, for apples, repeat at the pink bud stage, again when 80 per cent of the petals have fallen and then again three weeks after petal fall. For pears, repeat with captan at the white bud, petal fall and fruitlet stages, then again every two weeks.

Apple capsids (*Plesiocoris rugicollis*) suck sap from the shoot tips and fruitlets. Their toxic saliva damages the plant tissues and causes the foliage to develop many small holes, while the fruits become marked with corky scabs or raised lumps. The variety 'James Grieve' is particularly susceptible. Capsid bugs are pale green insects up to $\frac{1}{5}$ in long and occur on the tree between mid-April and July. Prevent damage by spraying at the green cluster stage with fenitrothion, dimethoate or formothion.

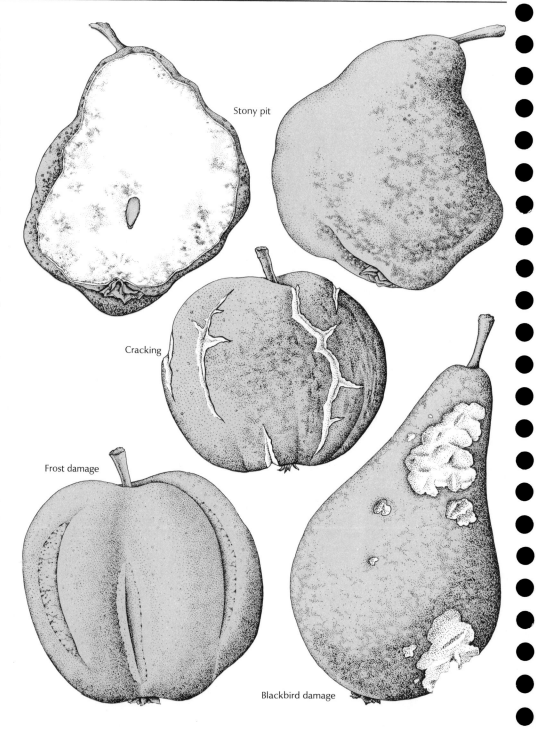

Stony pit

Cracking

Frost damage

Blackbird damage

Tree fruit: fruit 2

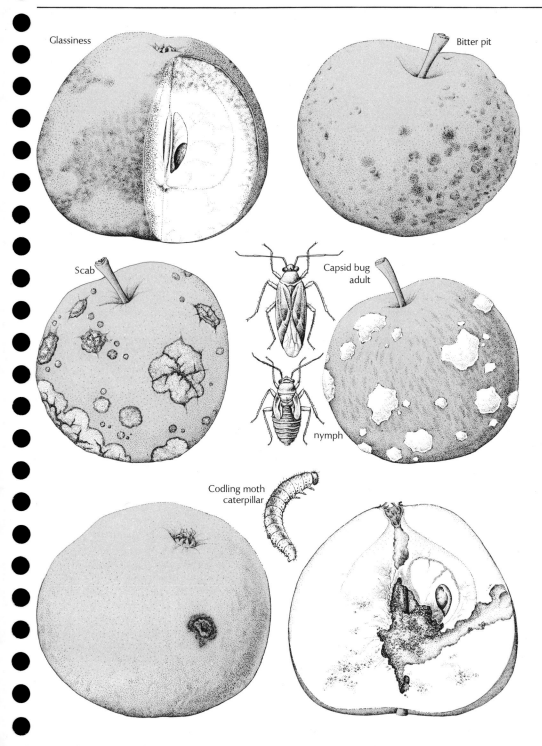

Glassiness

Bitter pit

Scab

Capsid bug adult

nymph

Codling moth caterpillar

Bitter pit of apples is caused by the fruit having either a deficiency of calcium or a too high concentration of potassium or magnesium. The symptoms are slightly sunken pits on the surface of the skin with small brown areas of tissue immediately beneath the pits and scattered throughout the flesh. Most pitting develops during storage, though some symptoms may appear while the fruit is still on the tree. If the brown patches occur near the skin they can be removed by peeling the fruit, but in severe cases the flesh has a bitter taste, thus making it inedible. Bitter pit appears to be connected with a shortage of water at critical times and can be prevented to a certain extent by watering during dry periods. Mulching also helps, but the use of straw can aggravate the trouble. Its incidence can be reduced greatly by spraying with 1 oz of calcium nitrate in 2 gal of water in mid-June, repeating ten days later, followed by a stronger solution of $1\frac{1}{2}$ oz calcium nitrate in 2 gal of water ten days later. Repeat the latter treatment at least three times at ten day intervals.

Glassiness or water core is a physiological disorder, the exact cause of which is not known. Large portions of the flesh from the skin right through to the core become waterlogged and affected flesh has a sweet or musty flavour. The trouble is most prevalent on the fruit of young trees just coming into bearing, particularly if these are growing in soil that is rich in nitrogen. It is worse in seasons when there are marked fluctuations in the weather conditions, and when days in late summer and early fall are hot, but the nights are cold with heavy dews. If the fruit is severely affected at the time it is picked, it may decompose in storage. Where fruits are only slightly affected, however, the glassiness may disappear in storage particularly if the fruits are kept in a cool store with adequate ventilation. Since the cause is not known, no preventive measures can be recommended, but because trees which are suffering from too dry soil conditions seem to be prone to the trouble, try to maintain even growth by mulching and watering in dry periods. It is particularly important that the soil around fruit trees is not allowed to dry out completely.

Mature fruits maggoty
Codling moth (*Laspeyresia pomonella*) attacks apples and, to a much lesser extent, pears. The moths lay their eggs in June and July, and the caterpillars tunnel into the maturing fruit. When fully fed in July or August the caterpillars leave the fruit and over-winter beneath loose flakes of bark. A related species, *Laspeyresia funebrana*, causes similar damage to plums. Control these pests by spraying the trees with any of the insecticides phosmet, phosalone or azinphos-methyl at two week intervals starting at petal fall. Some over-wintering caterpillars can be trapped by scraping loose bark from the trunk and larger branches in late July. Cover these cleared areas with sacking or corrugated cardboard. Then burn or boil this in November to kill the over-wintering caterpillars which have sought shelter under the coverings. Caterpillar trapping is only worth while on apple trees that are fairly isolated since untreated trees in nearby gardens will reinfest the area.

Fruitlets maggoty
Apple sawfly (*Rhagoletis pomonella*) lays its eggs on the flowers and the maggots attack the young fruitlets. Initially, the maggots tunnel just below the surface of the fruit causing a ribbon scar on the skin. Later they tunnel directly to the core of the apple. Fruitlets that have been damaged in this way fall from the tree in June or early July. Sometimes the maggots die before they can tunnel to the core, in which case the apple grows to maturity but is disfigured by the ribbon scar. Most cooking apples are not troubled by this pest but some dessert varieties such as 'Worcester Pearmain', 'Charles Ross', 'James Grieve' and 'Ellison's Orange' are particularly susceptible. Spray six weeks after petal fall, then at two week intervals, with phosmet, phosalone or azino-methyl.
Plum sawfly (*Rhagoletis flava*) tunnels into plum fruitlets but unlike the above species does not cause any distinctive scarring on the skin. Damaged fruitlets fall prematurely. The variety 'Czar' is frequently attacked while 'Pond's Seedling' and 'Monarch' escape damage. Kill newly hatched caterpillars by spraying with fenitrothion or dimethoate seven to ten days after petal fall.

Tree fruit: fruit 3

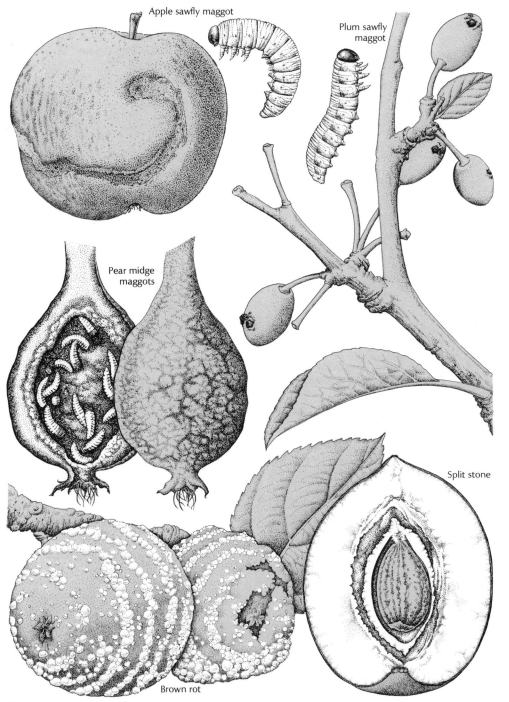

Apple sawfly maggot

Plum sawfly maggot

Pear midge maggots

Split stone

Brown rot

Pear midge (*Contarinia pyrivora*) lays its eggs in the unopened flower buds. Attacked fruitlets grow more rapidly than usual but a few weeks after petal fall they start to turn black and fall from the tree. If they are broken open, a dozen or more orange-white maggots about $\frac{1}{8}$ in long may be found inside. These maggots may already have gone into the soil to pupate, in which case the inside of the blackened fruitlets will be hollow. This pest is common in some localities and sometimes destroys most of the fruitlets. Spraying with fenitrothion when the blossom is at the white bud stage gives some protection. On small trees pick and burn infested fruitlets before the maggots leave to pupate.

Fruits rotting

Brown rot (*Monilinia fructicola* and *Monilinia laxa*) can infect all tree fruits. The fungus enters through wounds made by birds, wasps or caterpillars. It causes the fruits to turn brown and the flesh to become soft and decay rapidly. Infected fruits also become covered with concentric rings of buff or gray cushions of fungus spores, and eventually shrivel and dry up, either while still on the tree or on the ground. The disease spreads by contact and infected fruit, whether on the tree, on the ground or in store, should be removed and destroyed as soon as it is noticed. The rotting of apples in store may be reduced by spraying with benomyl in mid-August and early September before picking. Do not store wet, bruised or damaged fruits, including those from which the stalk has been torn off during picking. Keeping the store clean helps to reduce incidence of the disease. To control brown rot of peach, spray at blossom time with sulfur, captan or benomyl. Repeat at 14 day intervals until ripening. The fungus can also enter shoots and cause small cankers and die-back, therefore cut off and burn all dead shoots.

Split stone of peaches is a physiological disorder caused by one or more adverse factors. Affected fruits have a deeper suture than normal and are cracked at the stalk end, the hole sometimes being big enough for earwigs to enter. The stone of such fruit is split in two and the kernel (if formed) rots. Peaches with this trouble fail to ripen

FRUIT SKIN RUSSETED

Some varieties of apple naturally have a rough (or russeted) skin but, when it occurs on smooth-skinned varieties, russeting becomes a disorder. It can be due to frost damage when the fruit is young, an early attack of powdery mildew or even chemical damage. Faulty root action caused by malnutrition or other adverse soil conditions may also result in russeting. No control measures are required apart from the prevention of mildew, good cultivation and careful use of chemicals.

normally, and rot as a result of infection by brown rot fungus or other secondary organisms. Split stone can be due to poor pollination, therefore pollinate the flowers by hand if there is a shortage of pollinating insects. Do this by passing from flower to flower a soft camel hair brush tied on the end of a short bamboo cane. In dry weather gently syringe the flowers of trees on walls with tepid water about noon each day to help fruit set. A lack of lime may also cause split stone, therefore lime acid soils in the fall to raise the pH to 6.7–7.0. General malnutrition or feeding after stoning has started may also result in split stone. The commonest cause, however, is an irregular supply of moisture in the soil. Mulch and water in dry periods to keep it damp.

Tree fruit: roots

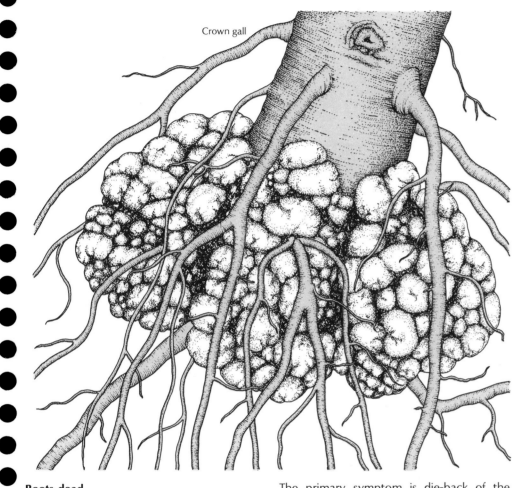

Crown gall

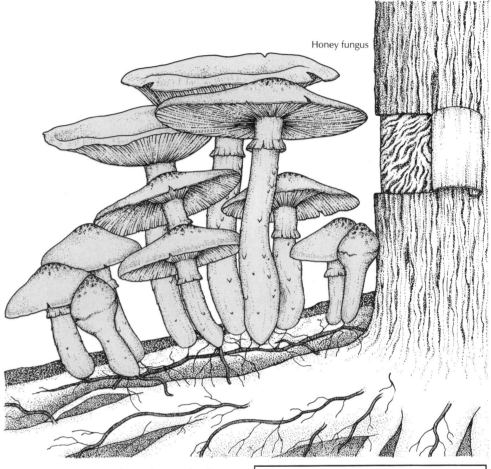

Honey fungus

Roots dead

Drought and waterlogging can both kill the finer roots of trees, resulting in discoloration of the foliage and sometimes die-back of the shoots. Whenever a tree dies back examine the roots to see if they are dead. Roots killed by drought show no distinctive character-istics, but those killed by waterlogging show a blue-black discoloration of the tissues, the outer layers of which peel off to leave a central core. Prevent such troubles by taking appropriate measures against physiological disorders, for example, mulching, watering or draining the soil (see pages 5–7).

Honey fungus (*Armillaria mellea*) is a serious disease that can cause trees to die suddenly.

The primary symptom is die-back of the stems (see under Stems dying back, page 13). Other symptoms are white fan-shaped growths of fungus appearing beneath the bark of the roots and trunk at or just above soil level, and brown-black root-like struc-tures known as rhizomorphs growing on diseased roots; these grow out through the soil to spread the disease. Another symptom is the appearance in fall of honey-colored toadstools at the base of a dying tree. Dig up and burn dead and dying trees together with as many of the roots as possible. Sterilize the soil with a 2 per cent solution of formalin applied at the rate of 6 gal per square yard, or with a proprietary phenolic emulsion.

Alternatively, change the soil to a depth of 2–3 ft before replanting.

Roots galled

Crown gall (*Agrobacterium tumefaciens*) can produce galls up to the size of footballs on the roots but the disease does not appear to affect the vigor of the tree or produce any symptoms on the leaves or shoots. The bacteria enter through wounds, and spread in wet soils. Improve the drainage should this trouble occur, and burn diseased roots. If a new tree is to be planted where an affected one has been removed, dip the roots in a copper fungicide before planting and try to avoid injuring them.

REPLANT DISORDER

Replant disorder is due to the fungus *Thielaviopsis basicola* on cherries, and on apples is believed to be due to a species of *Pythium*. On other fruits the cause is not known. It results in poor stunted growth or even death of the fruit tree if it is planted in soil where a tree of the same type had been growing previously for a number of years. When removing a fruit tree, either plant one of a completely different sort on the site or sterilize the soil with a 2 per cent solution of formalin applied at the rate of 6 gal per square yard.

Bush fruit 1

Leaves eaten

Gooseberry sawfly caterpillars (*Nematus ribesii*) feed on the leaves of gooseberry and red and white currant. Other related species also occur on these plants and on black currant. These caterpillars are pale green with black spots and feed in groups, especially during the early stages of their growth. By the time they are fully fed the bush may be entirely defoliated. The fruit is not damaged but the loss of leaves reduces the following year's crop. Several generations of caterpillars can occur between early May and September; check the plants regularly for their presence. Eggs are often laid low down or in the center of the bush and this is where the first signs of damage are likely to be found. Control these caterpillars by spraying with malathion, fenitrothion or derris. Use only derris if the fruit is almost ready for picking.

Magpie moth caterpillars (*Abraxus grossulariata*) eat the leaves of bush fruit between April and late May. The caterpillars are white with black markings and have a yellow stripe down their sides. They walk with a looping action and have only two pairs of clasping legs, which lie at the rear end of the body, unlike sawfly caterpillars which have seven pairs more evenly distributed. Magpie moth caterpillars are a pest only in certain parts of the United States. Control them by adopting the measures recommended for sawfly caterpillars (see above).

Leaves distorted

Aphids (various species) attack currant and gooseberry leaves at the shoot tips. One of the most common is the currant blister aphid (*Cryptomyzus ribis*), which causes raised blisters to appear on the leaves. Red and white currant leaves often turn red where the aphid has been feeding while black currant leaves show the same blistering with a yellow-green color. Other species cause varying degrees of leaf curling and stunting of the shoots. Honeydew excreted by the aphids makes the foliage and fruit sticky and encourages the growth of black sooty molds. The aphids over-winter on the bushes as eggs. Destroy them by spraying bushes with dormant oil in December or January. Follow this

with applications of a systemic insecticide such as dimethoate or formothion as soon as flowering has finished.

Capsid bugs (*Lygocoris pabulinus*) are pale green, sap-feeding insects which grow up to $\frac{1}{5}$ in long and attack the young leaves and shoot tips. Their toxic saliva causes small dead patches to appear on the leaves where the bugs have been feeding. As the leaves expand these dead areas tear to form numerous holes, and the foliage at the shoot tips develops a tattered and distorted appearance. The control measures given above for aphids are also effective in preventing capsid bug damage.

Leaves discolored

Silver leaf (*Stereum purpureum*) may affect red and black currants, producing a silvering of the leaves on one or more branches which later die back. On larger branches a brown or purple stain can be seen in the inner tissues. Cut out these branches to a point several inches below where the stain ceases and cover the wound with a wound paint. Feed, mulch and water as necessary to encourage vigor in the affected bushes and help them overcome this disease.

Faulty root action due to adverse soil conditions can result in discoloration of the foliage. Whenever leaves show discoloration consult the section on physiological disorders (see pages 5–7).

Leaves spotted

Leaf spot (*Pseudopeziza ribis*) affects currant and gooseberry leaves from May onwards. Small dark brown spots develop and coalesce until the leaf surfaces become totally brown. Premature leaf-fall occurs and the infected bushes lose vigor. During the winter the fungus remains dormant on the fallen leaves, so rake these up and burn them. Spray with zineb or thiram after flowering and repeat at 10–14 day intervals to within a month of fruit picking. Alternatively, spray with benomyl as the flowers open and repeat the treatment three times at 14 day intervals. Any of these fungicides can also be used after harvest if necessary. Feed diseased bushes well the following season to help them overcome the loss of vigor due to the early leaf-fall.

Gooseberry sawfly caterpillars

Capsid bug

Currant blister aphid

Bush fruit 2

Stems dying back

Honey fungus (*Armillaria mellea*) is a serious disease that can cause currant and gooseberry bushes to die suddenly. The leaves on the dead stems do not fall immediately but continue hanging for some time in a withered condition. Fan-shaped growths of fungus appear underneath the bark of the roots and stem at or just above soil level, and brown-black root-like structures known as rhizomorphs may be found on diseased roots; these grow out through the soil and spread the disease. In the fall honey-colored toadstools may appear at the base of the dying bush. Dig up and burn dead or dying bushes together with as many of the roots as possible. Sterilize the soil with a 2 per cent solution of formalin applied at the rate of 6 gal per square yard, or with a proprietary phenolic emulsion. Alternatively change the soil completely to a depth of at least 2 ft before replanting.

Faulty root action caused by adverse soil conditions can result in die-back of shoots. If no disease is apparent consult the section on physiological disorders, pages 5–7.

Stems with visible fungal growth

Gray mold (*Botrytis cinerea*) causes gooseberry bushes to die back branch by branch and eventually may kill the whole plant. The bark develops irregular cracks in which the fruiting bodies of the fungus appear as gray velvety cushion-like pustules of spores. These may also arise from hard black masses of fungal threads which are embedded in the bark. Cut out infected shoots to a point several inches below obviously diseased tissues and burn them. In severe cases the whole bush may have to be destroyed.

Coral spot (*Nectria cinnabarina*) can be very troublesome on red currant. The fungus enters through small wounds such as pruning snags and causes die-back of branches and sometimes the complete death of the plant. Numerous coral-red cushion-like pustules of spores develop on the dead shoots. Cut out all affected shoots to a point several inches below the diseased tissues and paint all wounds thoroughly with a fungicidal paint. The fungus can survive on dead wood, therefore destroy all woody debris such as old pea sticks.

Stems with pests visible

Brown scale (*Parthenolecanium corni*) occasionally occurs on old currant and gooseberry bushes. It is a sap-feeding insect that is covered by a shiny brown convex shell or scale up to $\frac{1}{4}$ in long. The scales are firmly attached to the older branches and do not move once they have found a suitable feeding place. Heavy infestations weaken the bush and reduce cropping. Control them by applying a dormant oil wash in December or January, or spray the bushes with malathion in early July with a second application two weeks later.

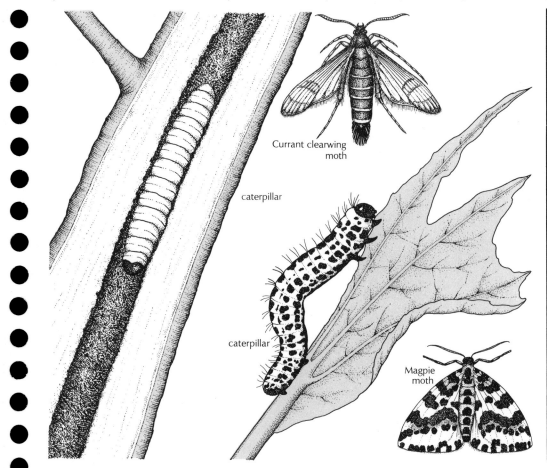

Currant clearwing moth

caterpillar

caterpillar

Magpie moth

GENERAL FUNGAL GROWTH

American gooseberry mildew (*Sphaerotheca mors-uvae*) affects both gooseberry and black currant. A white powdery coating of spores appears on the young leaves, shoots and especially the fruits. Later these patches turn brown and felt-like, and badly affected shoots become distorted at the tips. Prune regularly to keep the bushes open since the disease is most troublesome on overcrowded bushes where there is little circulation of air. Do not apply heavy doses of a nitrogenous fertilizer because this encourages soft growth which is susceptible to infection. Cut out and burn diseased shoots in late August or September. Spray with dinocap before flowering and repeat as necessary, or use benomyl or thiophanate-methyl as the first flowers open and repeat three times at 14 day intervals.

Bush fruit 3

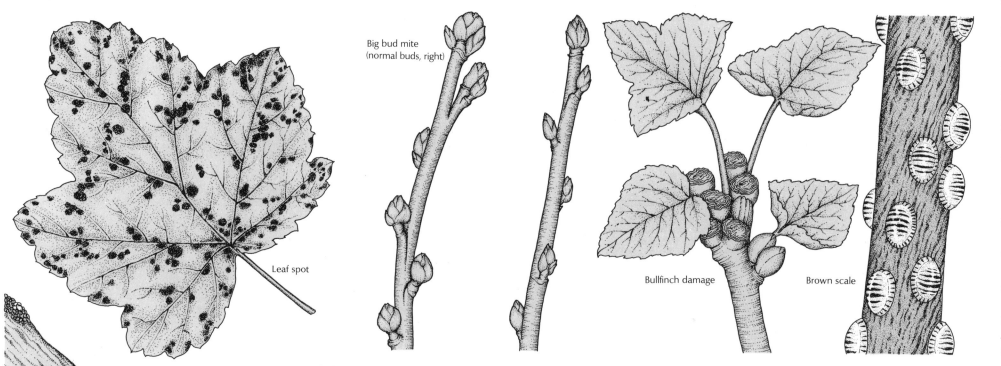

Leaf spot

Big bud mite (normal buds, right)

Bullfinch damage

Brown scale

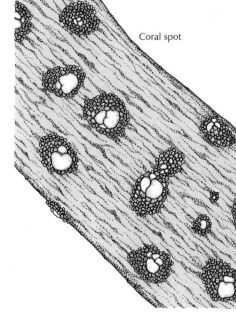

Coral spot

Stems tunneled
Currant clearwing moths (*Synanthedon salmachus*) attack mainly black currant but are also sometimes found on other currants and gooseberry. The moths have transparent wings and yellow stripes on their black abdomen, which cause them to resemble small wasps. They lay their eggs on sunny days in June. These hatch into white caterpillars with brown heads, which tunnel into the younger stems. The damaged stems usually survive but they are weakened and may snap if carrying a heavy crop. Attacks are not usually severe enough to require the use of insecticides. Instead prune out infested shoots during the winter and burn them. These shoots can be detected by their tendency to snap when bent slightly.

Buds eaten
Bullfinches (*Pyrrhula pyrrhula*) are capable of eating most of the flower buds of gooseberry during the winter. In severe attacks only a few buds at the tips of the shoots escape, resulting in a crop of only a handful of fruits.

Currants are less susceptible, but black currant flower buds may be eaten after bud burst. In areas where bullfinches are a problem grow the bushes under a fruit cage, or net them before damage starts. Bird repellant sprays do not give reliable control when alternative sources of food are scarce.

Buds enlarged
Big bud mites (*Cecidophyopsis ribis*) are tiny mites that live in large numbers inside the buds of black currant, the only bush fruit that they attack. Infested buds are readily detected, especially during the winter, because they are swollen and round—healthy buds being pointed and smaller. In the spring infested buds either fail to grow or produce a few distorted leaves. The mites emerge at this time and seek out new buds to invade. In addition to the direct damage caused by the mite, it is also important as a carrier of reversion disease (see Box). There are no really effective control measures for this pest available to amateur gardeners. On lightly infested bushes check the spread of mites by picking

off and burning enlarged buds during the winter. Spraying with benomyl when the first flowers open, with two further applications at 14 day intervals, also checks the mites. Scrap heavily infested bushes and, if possible, grow the replacements in a different part of the garden.

Flowers discolored
Frost damage when the bushes are in bloom causes browning of the flowers and subsequent loss of crop. Covering the bushes with newspaper or net curtains when a frost is forecast may help to prevent such damage.

Fruits lacking
Various troubles may cause fruit to be lacking. These include bird damage, poor pollination, frost damage, reversion (see Box) and malnutrition, especially potassium deficiency in currants. Most of these troubles will also have produced other symptoms earlier in the season, so keep a watch on bushes throughout the growing season, and carry out preventive treatments where appropriate.

Bush fruit 4

Gray mold

Honey fungus
on root

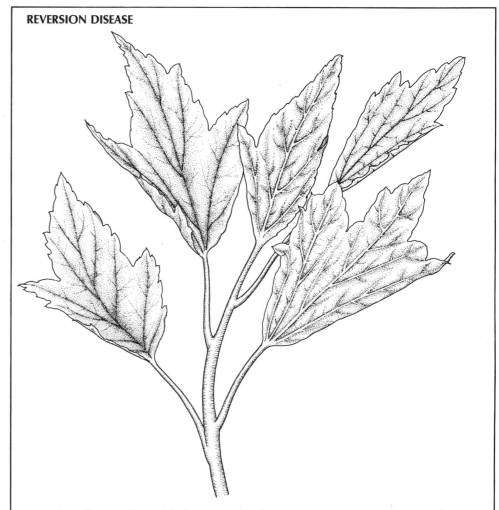

REVERSION DISEASE

Fruits blemished
Scald is caused by the sun's rays striking moist fruits and occasionally occurs on gooseberries. A white depressed patch may show on the side of the fruit, or they may resemble boiled gooseberries. Remove all affected fruits.
Splitting of fruits is usually due to irregular growth caused by faulty root action. Try to keep growth even by mulching and watering in dry periods before the soil dries out completely. Remove affected fruits before they are attacked by secondary organisms.

Fruits with visible fungal growth
Gray mold (*Botrytis cinerea*) can cause serious losses of currants in a wet summer. The disease shows as a gray-brown fluffy growth covering the currants. Although the symptoms are not usually seen until the fruit is reaching maturity, the fungus actually enters at flowering time. Therefore prevent infection by spraying with benomyl or thiophanate-methyl as soon as the flowers open and repeat three times at 14 day intervals, or

apply dichlofluanid when the flowers open and repeat four times at ten day intervals. Remove infected strigs of currants since the disease spreads rapidly by contact.

Roots dead
Honey fungus (*Armillaria mellea*) attacks the roots, and frequently causes currant and gooseberry bushes to die suddenly. For symptoms and treatment of honey fungus, see under Stems dying back, page 19.
Drought and waterlogging can both kill the finer roots of currant and gooseberry bushes. This produces a discoloration of the foliage and sometimes die-back of the shoots. Whenever a bush dies back examine the roots to see if they are dead. Roots killed by waterlogging show a blue-black discoloration of the tissues, the outer layers of which peel off to leave a central core. If the roots were killed by drought then there are no distinctive symptoms. Prevent such troubles by taking appropriate measures against physiological disorders, for example mulching, watering or draining the soil (see pages 5–7).

Reversion disease affects black currants and is due to a viral-like organism that is spread by big bud mites. In June or July the affected plants develop abnormal leaves on the basal shoots. When mature these leaves are narrow and have less than five pairs of veins on the main lobe. These symptoms are, however, difficult for an amateur to identify, particularly in spring and late summer when many of the young leaves on healthy bushes may look abnormal. Therefore the disease is better diagnosed by examining the flower buds.

These turn bright magenta instead of the usual dull gray color of the flower buds on healthy bushes. If bushes fail to crop properly look for this symptom, and also the distortion of the foliage described above, the following season—particularly if they were also affected by big bud mite. Destroy diseased bushes, and plant new bushes, of stock certified to be disease-free, on a fresh site. Prevent the disease from occurring by controlling big bud mites. For a description of this pest and its control, see under Buds enlarged.

Cane fruit 1

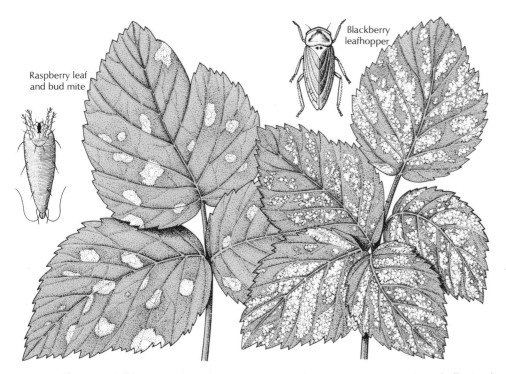

Raspberry leaf and bud mite

Blackberry leafhopper

Aphids

Lime-induced chlorosis on raspberry

Leaves with pests visible

Aphids (several species) infest the leaves, shoot tips and flowers of cane fruits during the summer. In addition to the direct damage caused by their sucking sap, aphids also transmit virus diseases. Check the spread of such diseases by spraying with a systemic insecticide in April when aphids start to become active on the plants. Further treatments may be necessary later in the summer. In winter apply a dormant oil to control over-wintering eggs. Some varieties of raspberry, such as 'Malling Delight', 'Malling Orion' and 'Malling Leo', are resistant to aphids.

Leaves discolored

Lime-induced chlorosis is a common problem on raspberry and is caused by the soil being too alkaline. It shows as a yellowing between the veins and the leaves may become almost white, especially at the tips of the spurs. Try to prevent this trouble by digging in plenty of humus, particularly acidic materials such as peat or crushed bracken. Do not add alkaline materials such as mushroom compost or nitro-chalk. Apply a chelated compound each season or rake in fritted trace elements.

Magnesium deficiency shows as orange bands between the veins, particularly on raspberry after a wet spring. At the first sign of trouble spray with a solution of magnesium sulfate at a rate of $\frac{1}{2}$ lb in 3 gal of water to which is added a spreader. Give two or three applications at two week intervals.

Manganese deficiency may occur on raspberry in wet seasons. It causes yellowing between the veins, and is difficult to distinguish from lime-induced chlorosis and magnesium deficiency. If other treatments have been carried out but have failed, spray with a solution of manganese sulfate, 2 oz in 3 gal of water plus a spreader, two or three times at two week intervals.

Faulty root action due to adverse soil conditions can result in discoloration of the foliage. Whenever leaves show discoloration which cannot be attributed to any disease of the canes consult the section on physiological disorders, pages 5–7.

Leaves spotted

Leafhoppers (*Ribautiana* spp) suck sap from the undersides of the leaves. The adults are pale yellow insects about $\frac{1}{8}$ in long, which readily jump off the leaves when disturbed. Their cream-white nymphs are less active and more easily seen. Small pale green or white dots appear on the upper leaf surfaces as a result of the leafhoppers' feeding. Eventually these dots coalesce and much of the leaves' green color may be lost. Like aphids, leafhoppers can transmit virus diseases. Control these pests by spraying with a systemic insecticide when symptoms are first noticed.

Leaf and bud mites (*Phytoptus gracilis*) are microscopic pests that live on the undersides of the leaves and cause yellow blotches to appear on the upper leaf surfaces. Since mosaic virus also causes this symptom the two can be confused but, unlike the virus, these mites do not have any adverse effect on the plant. This is fortunate as there are no chemicals currently available to amateur gardeners that can control this mite. While the plants continue to crop normally it can be assumed that mosaic virus is not present and the disfigurement of the foliage caused by the mite can be tolerated. The mite is more common than mosaic virus, especially on raspberry and, to a lesser extent, on blackberry and hybrid berries.

Cane spot (*Elsinoe veneta*) attacks raspberry, loganberry and hybrid berries, but is rare on blackberry. It shows as very small purple spots on the leaves, a symptom that is only likely to occur if the canes are also affected. For treatment see under Canes spotted.

Canes galled

Crown gall (*Agrobacterium tumefaciens*) causes raspberries and blackberries to develop a walnut-sized gall at ground level or a chain of small galls higher up the canes. The bacteria enter through wounds, and spread in wet soils. Burn severely infected canes and replant with disease-free canes on a fresh site where the drainage is better. Help to prevent this disease by ensuring that plants are not over-watered.

Cane fruit 2

Raspberry spur blight

Aerial crown gall on blackberry

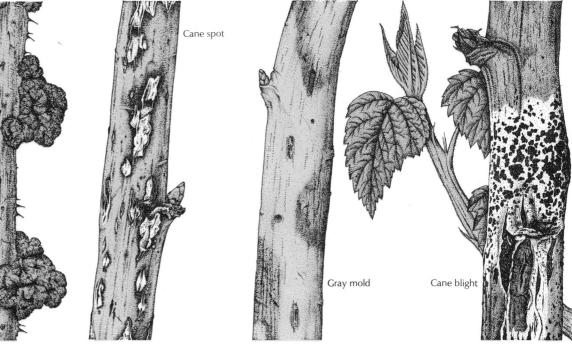

Cane spot

Gray mold

Cane blight

Canes spotted

Cane spot (*Elsinoe veneta*) affects raspberry, loganberry and hybrid berries, but only rarely blackberry. It first appears in May or June as small purple spots on the canes. As the spots grow to their maximum size of $\frac{1}{4}$ in long, they become elliptical and pale with a purple border. Eventually they split to form shallow pits or small cankers. In severe cases the tips of the shoots die back. Symptoms may also develop on the leaves and fruit, the leaves developing small purple spots and fruits becoming distorted. The fungus that causes the disease over-winters on the canes and reinfection takes place from May until October. Cut out and burn badly spotted canes in the fall. For raspberries spray with liquid lime sulfur or captan at bud burst and just before flowering. An alternative is to use benomyl or thiophanate-methyl at bud burst and repeat every two weeks until the end of flowering. For loganberries spray with bordeaux mixture, liquid lime sulfur or captan just before flowering and when the fruit sets.

Canes dying back

Spur blight (*Didymella applanata*) is one of the most troublesome diseases of raspberry, and it can also attack loganberry. The fungus infects new canes in June, particularly those which are overcrowded, but the symptoms do not usually appear until August when dark purple blotches develop around the nodes. The blotches enlarge, turn silver and then become studded with minute black fruiting bodies of the fungus. The buds at the affected nodes either die or produce shoots in the spring that soon die back. Remove superfluous canes early and cut out and burn diseased canes at the first signs of the trouble. Spray with benomyl, dichlofluanid or captan when the new canes are a few inches high and repeat three times at 14 day intervals, or spray with thiophanate-methyl at bud burst and repeat every 14 days until the end of flowering. Alternatively, spray affected canes with bordeaux mixture, or any other copper fungicide, at bud burst and repeat the treatment when the flower tips are just showing white.

Cane blight (*Leptosphaeria coniothyrium*) is a fungal disease that affects only raspberries, particularly the varieties 'Norfolk Giant' and 'Lloyd George'. The fungus enters the canes at the base, usually through frost cracks. The canes then develop a dark area just above ground level and become so brittle that they can be snapped off easily. The leaves on infected fruiting canes wither during the summer. The soil can become contaminated with the fungus once it has entered canes at soil level. Cut diseased canes hard back to below soil level and burn them. Disinfect the pruning tools immediately afterwards. Spray newly developing canes with either of the fungicides captan or folpet.

Gray mold (*Botrytis cinerea*) can kill raspberry canes in the winter and early spring. The fungus usually enters through frost cracks and produces hard black resting bodies up to $\frac{1}{8}$ in long. These are embedded in the tissues of the dead canes, which become either silvery or paler in color. In wet weather gray velvety cushion-like spores of the fungus may arise from the resting bodies. Cut out

infected canes and burn them. Note that the fungus can spread to the fruits later in the season. For symptoms and treatment, see under Fruits rotting.

Honey fungus (*Armillaria mellea*) often causes the rapid death of canes. The leaves do not, however, fall immediately but continue hanging for some time in a withered condition. White fan-shaped growths of fungus develop beneath the bark of the roots and around the crown of an infected plant, and dark brown root-like structures known as rhizomorphs may be found growing in the diseased soil; these grow out through the soil and infect new plants growing nearby. Dig up and burn diseased stools together with all the roots. Sterilize the soil with a 2 per cent solution of formalin (made by dissolving 1pt of formalin in 6 gal of water), applying this solution at the rate of 6 gal per square yard of the infected area.

Faulty root action due to adverse soil conditions can result in die-back of the canes. If no disease is apparent consult the section on physiological disorders, pages 5–7.

Cane fruit 3

Raspberry beetle adult

larva

Gray mold on raspberry

Cane spot on raspberry

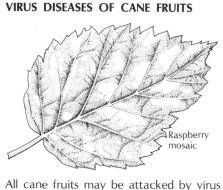

VIRUS DISEASES OF CANE FRUITS

Raspberry mosaic

All cane fruits may be attacked by virus diseases but raspberry is particularly susceptible. Viruses that affect raspberry include raspberry yellow net, leaf spot and leaf curl viruses. They are all aphid-borne and are responsible for the disease known as raspberry mosaic. Other viruses are either transmitted by eelworms in the soil, for example arabis mosaic virus, or are seed- or pollen-borne. In general the symptoms of virus infection are yellow mottling or blotching of the leaves, with distortion. Growth is poor and yield is reduced on severely affected canes.

Raspberry, hybrid berries and especially blackberry can also be affected by a mycoplasma—an organism that has certain characteristics in common with both viruses and bacteria. The disease it causes is known as stunt and is very similar to a virus disease. Infected plants are extremely stunted and the crop is reduced since few flowers are produced. The disease is transmitted by leafhoppers.

There is no cure for virus or mycoplasma infection and affected stools should be removed and burnt. Buy only canes certified to be healthy and replant on a fresh site at least 50 ft away from any old canes which are to be retained. If a fresh site is not available, destroy all the old canes at the same time and change the soil completely to a depth of 1½ ft over an area 2 ft wide before replanting. If possible plant cane fruits well away from hedges since they may harbor eelworm carriers of some of the viruses.

Fruits distorted

Poor pollination can result in distorted fruits which may show a small flattened or brown area. This trouble often occurs when bad weather conditions at flowering time discourage the pollinating insects. There is nothing that can be done to prevent it.

Cane spot (*Elsinoe veneta*) can cause distortion of the fruits on canes that are severely affected by this disease. The symptoms on the leaves and canes (see above) would, however, be more obvious than those on the fruit. Prevent the trouble by spraying as recommended above under Canes spotted.

Fruits maggoty

Raspberry beetles (*Byturus tomentosus*) are small brown beetles that lay their eggs on the flowers of raspberry, blackberry and loganberry. These hatch into pale brown larvae which feed on the ripening fruits, especially at the stalk end, and cause the damaged parts to dry up. Control this pest by spraying with derris or malathion. For raspberry, spray when the first pink fruit develops, for logan-

berry at 80 per cent petal fall and again two weeks later, and for blackberry spray when the first flowers open. Spraying during full flower must be avoided otherwise pollinating insects will be killed. Even at the times recommended above there is some risk. Therefore apply insecticides at dusk when bees have returned to their hives.

Fruits rotting

Gray mold (*Botrytis cinerea*) can cause considerable rotting of berries in a wet season. The affected berries become covered with a gray-brown fluffy growth of the fungus. If possible remove and destroy infected fruits to prevent spread of the disease by contact and by air-borne spores. Spray with benomyl or thiophanate-methyl as soon as the flowers open and repeat three times at 14 day intervals. Note that too frequent use of these fungicides can lead to the build-up of tolerant strains of the fungus. Alternatively spray affected canes with dichlofluanid as the flowers open, and repeat the treatment four times at ten day intervals.

Roots dead

Waterlogging, or less frequently drought, can kill the roots of cane fruits, resulting in discoloration of the foliage and sometimes die-back of the canes. Roots killed by waterlogging have a blue-black discoloration and the outer tissues peel off leaving a central core. Prevent such troubles by appropriate remedial measures such as mulching, watering or draining the soil (see the section on physiological disorders, pages 5–7). If cane fruits are to be planted in heavy clay soil, try to lighten it by digging in plenty of humus over the whole area, and raise the bed about 6 in as this will help it to drain. Canes showing signs of ill health due to adverse soil conditions may recover following applications of a foliar feed.

Honey fungus (*Armillaria mellea*) attacks both the roots and crowns of all types of cane fruit. Affected roots develop dark brown cords known as rhizomorphs which spread out through the soil to infect plants nearby. For further symptoms and treatment see under Canes dying back.

Strawberries 1

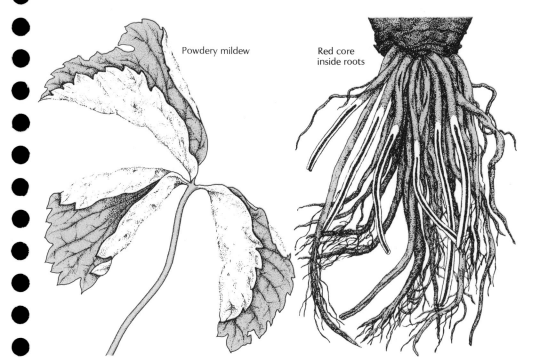

Powdery mildew

Red core inside roots

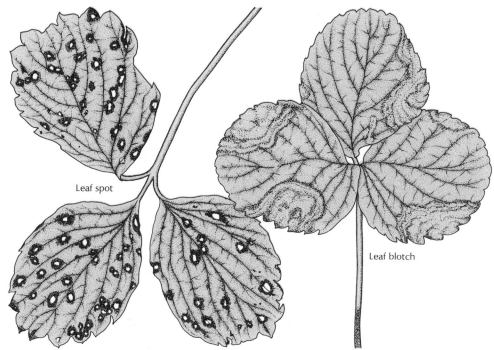

Leaf spot

Leaf blotch

Leaves with pests visible
Strawberry aphids (*Chaetosiphon fragaefolii*) and other species of aphid suck sap from the foliage during the summer. Heavy infestations check growth, reduce cropping and may spread virus diseases throughout the strawberry bed. Control aphids by spraying with a systemic insecticide in late April before the flowers open. Further treatments may be needed later in the summer after the crop has been gathered.

Leaves discolored
Red spider mites (*Tetranychus urticae*) may cause considerable damage in hot dry summers. The tiny yellow-green mites occur in large numbers on the undersides of the leaves where they suck sap. This causes a pale mottled discoloration and the leaves may eventually turn yellow, dry up and die. At this stage a fine silken webbing may be visible on the plants. Control the mites by spraying the undersides of the leaves thoroughly with malathion or a systemic insecticide, with two further applications at

seven day intervals. Prevent damage by treating when the symptoms first appear.
Powdery mildew (*Sphaerotheca macularis*) causes the leaves to turn purple and curl upwards. The undersides of the leaves become gray due to the development of white fungal spores by which the disease spreads. The fungus over-winters on old leaves. Cut them off after harvesting. Dust with sulfur or spray with dinocap just before flowering and repeat every 10–14 days until two weeks before harvest. Alternatively, spray with benomyl or thiophanate-methyl just after flowering, repeating two and four weeks later. Any of the above fungicides may also be applied after harvesting as an alternative to removing the old leaves. 'Royal Sovereign' and 'Cambridge Vigour' are very susceptible to powdery mildew but 'Cambridge Favourite' is resistant.
Lime-induced chlorosis is caused by the soil being too alkaline. This results in the development of yellow bands between the veins of the leaves, which may later become almost white. Try to prevent this trouble by digging

in plenty of humus, particularly acidic materials such as peat, crushed or chopped bracken, or pulverized tree bark. Do not add alkaline materials, for example mushroom compost or nitro-chalk. Rake in fritted trace elements or use a chelated compound each season.
Faulty root action may be caused by drought, waterlogging, or poor planting, that is, the roots not being well spread out during planting. They can all result in discoloration of the foliage. Prevent such troubles by careful planting and by suitable remedial measures (see the section on physiological disorders, pages 5–7). It is usually possible to restore the vigor to affected plants by spraying with a foliar feed during the growing season.

Leaves spotted
Strawberry leaf spot is caused by the fungi *Diplocarpon earliana* and *Mycosphaerella fragariae* and, rarely, *Stagonospora fragariae*. They are usually found only on older leaves or those of plants lacking in vigor, and show as small red or purple spots which sometimes

become gray or brown with a red or purple margin. Occasionally, severe infections occur causing the spots to coalesce and the leaves to wither and disintegrate. Remove and burn any spotted leaves. The following season spray developing leaves with a foliar feed to encourage vigor. If the trouble persists spray with captan, thiram or benomyl, applying it in spring just after growth starts. Repeat applications of fungicide at ten day intervals until three days before the strawberries are to be harvested.
Strawberry leaf blotch (*Gnomonia fragariae*) has variable symptoms. It may show as large brown blotches with a purple border and surrounded by a zone of yellow. Alternatively the leaf and flower stalks may blacken and rot with subsequent death of the leaves and withering of the fruits. Remove and burn infected leaves and all dead leaf and flower stalks. The following season spray all strawberry plants with captan, thiram or benomyl in spring just after growth starts. Repeat applications of fungicide until three days before the crops are to be harvested.

Strawberries 2

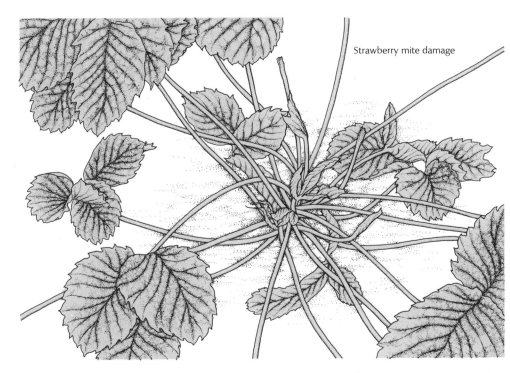

Strawberry mite damage

Black eye

Strawberry seed beetle

Strawberry green petal

Crowns withering

Faulty root action is caused by drought, waterlogging or poor planting and can result in the withering of all the top growth of affected plants. Prevent such troubles by careful planting and by the measures given in the section on physiological disorders (see pages 5–7).

Soil-borne diseases may be caused by any of the fungi *Thielaviopsis basicola* (black root rot), *Verticillium* spp (verticillium wilt) and *Phytophthora fragariae* (red core). They can all result in withering of the foliage or even the complete death of plants. In each case the roots will be dead or discolored either externally or internally, but these troubles can only be identified positively by microscopic examination and specialist advice should be sought if plants die for no obvious reason. Infected plants must be burned. After determining the exact cause of the disease, plant strawberry varieties that are listed as being resistant to that disease. Do not plant strawberries in the year following tomatoes.

Crowns stunted

Eelworms and strawberry mites (*Aphelenchoides* spp and *Tarsonemus pallidus*) are both microscopic pests that live inside the crowns of strawberries and cause the leaves to become distorted and undersized, and the plant to be stunted. An infestation may start with just one or two plants being attacked, but gradually the trouble will spread throughout the bed. There are no chemicals available to amateur gardeners for controlling either of these pests. Therefore, remove and burn stunted plants. When replanting use only certified stock, which should be grown in a different part of the garden.

Flowers discolored

Black eye is due to frost damage causing the central portion of affected flowers to turn black. The petals, however, remain unharmed. The flowers then wither and fall, by which time there is no effective treatment. Prevent this trouble by protecting plants with cloches, old net curtains or newspaper if frost is forecast at flowering time.

Strawberry green petal is caused by a mycoplasma, an organism intermediate between viruses and bacteria. The flowers of an affected plant have enlarged sepals but small green petals, and they usually shrivel up and fail to produce edible fruits. The young leaves of infected plants are very small and yellow, later turning red and wilting. The whole plant may collapse and die in midsummer, although there may be a temporary recovery in September. The disease is spread by leafhoppers so control these pests by spraying with a systemic insecticide. A frequent source of infection of this disease is clover so keep strawberry crops free of such weeds. Destroy all infected plants.

Fruits eaten

Strawberry seed beetles (*Harpalus rufipes*) are active, shiny black beetles about $\frac{5}{8}$ in long. They feed on weed seeds and also the seeds on the outside of strawberries, which become discolored where the seeds have been removed. Damage begins before the fruits are ripe. Discourage these beetles by keeping the strawberry bed free of weeds throughout the year. If beetles are present catch them in pit-fall traps by sinking jam jars into the soil up to their rims. Methiocarb slug pellets are also effective against the beetles.

Slugs and birds both eat the ripening fruits. Control slugs by scattering methiocarb pellets among the plants before and after strawing down. Birds can eat the entire crop; protect against them by covering the strawberry bed with netting before the fruit is ripe. If possible, remove the netting from the bed during the flowering period in order to give pollinating insects better access to the flowers.

Fruits lacking

The commonest causes of poor cropping are strawberry leaf blotch, virus infection and black eye. For the symptoms and treatment, see the relevant entries above.

Fruits distorted

Poor pollination can result in distorted fruits since, where seeds do not develop because of the lack of fertilization, the fruits fail to

Strawberries 3

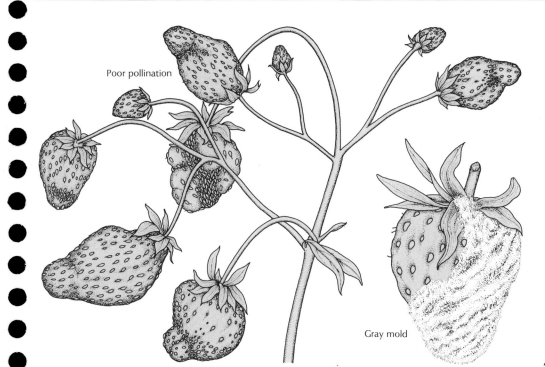

Poor pollination

Gray mold

Strawberry root weevil

adult

grub

swell. This trouble often occurs when bad weather at flowering time discourages pollinating insects. There is nothing that can be done to prevent it.

Fruits rotting

Gray mold (*Botrytis cinerea*) causes the berries to rot and become covered with a brown-gray fluffy growth of fungus. The fungus enters the flowers but the disease does not usually show until the fruit is beginning to ripen. Once it appears it may spread rapidly by contact between diseased and healthy fruits and, in a wet summer, serious losses of fruit can occur. By the time disease shows on the fruit it is too late to apply fungicides. Remove and destroy all infected fruits when harvesting to reduce the risk of infection. To prevent infection spray as the first flowers open with dichlofluanid and repeat four times at ten day intervals. Alternatively, apply benomyl or thiophanate-methyl as soon as the flowers open and repeat two, four and six weeks later. Note that regular use of these fungicides could lead to

the build-up of tolerant strains of the fungus so that the fungicides cease to be effective. Thiram and captan can also be used to control gray mold except on fruit that will later be preserved. There are always spores of gray mold in the air and it is frequently found on dying weeds, and also on thick straw mulches. Therefore keep weeds under control and do not mulch too heavily around strawberry plants.

Roots eaten

Strawberry root weevil grubs (*Otiorhynchus ovatus*) are plump, white, legless grubs with brown heads. They grow up to $\frac{1}{2}$ in long and feed on the roots and burrow up into the crown of the plant. Infestations are usually not noticed until most of the roots have been destroyed and the plant is wilting. Carefully dig up and burn damaged plants to destroy as many of the grubs as possible. Protect the remaining plants by spraying them with a solution of carbaryl or malathion from late in the blossom stage until three days before the plants are harvested.

VIRUS DISEASES OF STRAWBERRIES

Strawberries can be affected by a number of virus diseases. These may be caused by one or more different viruses and are usually spread by aphids. In general the symptoms are stunting or the complete collapse of a plant, with little if any fruit developing. The symptoms are most obvious in April and September when diseased plants show leaf symptoms such as yellow edges, yellow or purple blotching, and dwarfing and puckering.

Destroy all plants showing any of these symptoms and do not take any runners from them. If a stock is severely affected burn all the plants. Buy only those plants that are certified to be healthy and if possible plant them on a fresh site well away from hedges. Wash hands and tools thoroughly with soapy water after removing diseased strawberry plants, otherwise the virus may subsequently spread to other parts of the garden.

Outdoor vines 1

This section is concerned only with vines grown out of doors. For greenhouse vines see Protected crops, pages 54–6.

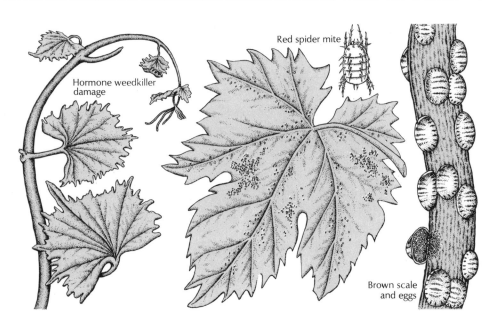

Hormone weedkiller damage

Red spider mite

Brown scale and eggs

Mealybugs

Shanking

Leaves distorted

Hormone weedkiller damage caused by the misuse of selective herbicides such as 2,4-D and dicamba is a very common trouble on vines. The leaves become narrow and fan-shaped, are frequently cupped, and the shoots twist spirally. Infected plants usually grow out of the symptoms in due course. Prevent such damage by the careful use of hormone weedkillers. They should be applied only on a still day using equipment kept specifically for their use.

Leaves discolored

Magnesium deficiency shows as a yellow-orange discoloration between the veins. In some varieties of vine, however, the blotches may be purple. In both cases the infected areas eventually turn brown. At the first signs of trouble spray with $\frac{1}{2}$ lb magnesium sulfate in 3 gal of water plus a spreader. Repeat once or twice at two week intervals.

Faulty root action is caused by drought, waterlogging or general malnutrition and can result in yellowing or browning of the leaves, or even die-back of shoots. Prevent such troubles by careful planting, correct cultural treatment including feeding, mulching and watering as necessary. If the soil is subject to

waterlogging, carry out appropriate remedial measures as given in the section on physiological disorders, pages 5–7. It is often possible to restore the vigor to affected plants by spraying with a foliar feed during the growing season.

Red spider mites (*Tetranychus urticae*) are tiny yellow-green creatures that suck sap from the foliage. Damage commences during May or June and shows initially as a fine mottled discoloration on the upper surface of the leaves. Later the leaves become yellow, dry up, and may fall off. Treatment is necessary as soon as symptoms are seen if damage is to be avoided. Thoroughly spray the foliage, especially on the undersides, with malathion or a systemic insecticide. At least three applications at seven day intervals will be needed to control these mites.

Stems with obvious pests

Scale insects such as brown scale (*Parthenolecanium corni*) and woolly vine scale (*Pulvinaria vitis*) encrust the bark on the older wood of vines. The former have red-brown oval shells which are about $\frac{1}{4}$ in long when fully grown. Woolly vine scales are the same size and have oval, wrinkled, brown shells which perch on the edge of white egg sacs.

Control the over-wintering stages of these scales by spraying or painting the vine rods with dormant oil in December, preferably after scraping off any loose bark that is present. Alternatively spray the vines with malathion in early July, giving a second application of malathion two weeks later when the young scales are hatching.

Mealybugs (*Pseudococcus* spp) are pinkish-gray, soft-bodied insects up to $\frac{1}{5}$ in long that suck sap from young stems and leaf axils (the point where the leaf-stalk meets the stem). They secrete a fluffy white wax with which they cover themselves and their eggs. In heavy infestations the foliage and fruits may become sticky with honeydew excretions upon which sooty mold may grow, especially in wet seasons. The winter treatment described above for scale insects also controls mealybugs. If the pest is present in the summer, control it by spraying vines with malathion.

Stems dying back

Honey fungus (*Armillaria mellea*) frequently kills vines. It causes stems to die back, and the leaves wither but continue to hang for a while before falling. White mats of fungal growth develop beneath the bark of the roots

and the main stems of the vine at or just above soil level, and dark brown root-like structures known as rhizomorphs develop on the diseased roots. These grow out through the soil to spread the disease to nearby woody plants. Dig up and burn dead and dying vines together with as many of the roots as possible. Sterilize the soil with a 2 per cent solution of formalin applied at the rate of 6 gal per square yard.

Root death due to drought or waterlogging causes die-back of shoots. For symptoms and treatment see below under Roots dead.

Fruits eaten

Wasps and birds eat the grapes as they ripen. Bird damage can be prevented by netting the vine. To prevent wasp damage enclose individual bunches of fruit in bags made of muslin or old nylons. Search for and destroy wasp nests by placing carbaryl dust in the entrance at dusk.

Fruits blemished

Splitting of berries is usually caused by powdery mildew (see Box), but when it occurs on otherwise healthy berries irregular watering is responsible. Prevent this trouble by mulching and watering before the soil dries

Outdoor vines 2

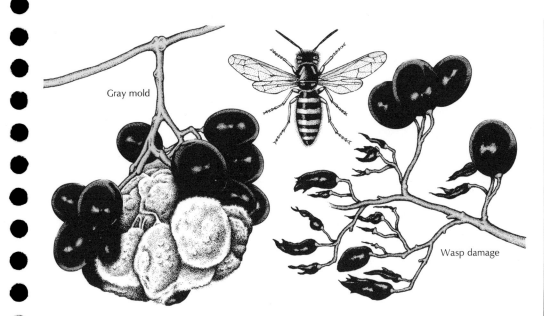

Gray mold

Wasp damage

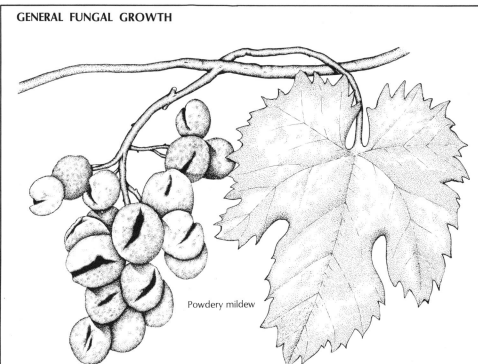

GENERAL FUNGAL GROWTH

Powdery mildew

out completely. If practicable remove the affected berries before they are attacked by secondary organisms such as gray mold.

Shanking is caused by certain unsuitable cultural conditions. The stalk of the berry shrivels gradually until it is completely girdled. At the early ripening stage occasional berries or groups of berries fail to color normally, those on black varieties turning red while those on white varieties remain translucent. The berries do not swell normally and are watery and sour. The trouble can be caused by drought or waterlogging, therefore carry out suitable remedial measures as recommended in the section on physiological disorders, pages 5–7. Overcropping puts an undue strain upon the root system so reduce the crop for a year or two until the vine has regained its vigor. When shanking occurs early in the season cut out the withered bunches and spray with a foliar feed.

Fruits rotting

Gray mold (*Botrytis cinerea*) affects the berries, causing them to rot and become covered with a gray-brown growth of fungus. Losses may be severe in a wet season. Infection usually occurs through the flowers but no symptoms appear until the berries

are nearly ripe. The disease then spreads rapidly by contact. Prevent infection by spraying when the first flowers open with dichlofluanid, benomyl or thiophanate-methyl, repeating twice at 10–14 day intervals. Note that regular use of the two latter fungicides could lead to the development of tolerant strains of the fungus, thus rendering these fungicides ineffective.

Roots dead

Drought and waterlogging can kill the finer roots of vines, causing discoloration of foliage or even die-back of the shoots. Whenever a vine starts to die back examine the roots to see if they are dead. Roots killed by drought show no distinctive characteristics except perhaps for slight shriveling, but those killed by waterlogging show a blue-black discoloration of the tissues, and the outer layers peel off leaving a core. Prevent such troubles by appropriate measures such as mulching, watering or draining the soil. See the section on physiological disorders, pages 5–7.

Honey fungus (*Armillaria mellea*) causes the roots to die and may kill the vine. For symptoms and treatment see under Stems dying back, page 28.

Powdery mildew (*Uncinula necator*) shows as a white floury coating of fungus spores on the leaves, young shoots and fruits. Affected berries drop if attacked early but in later attacks become hard, distorted and split and are then infected by secondary fungi such as gray mold (see Fruits rotting). Since the disease is encouraged by humid atmospheres prune carefully to avoid overcrowding of the shoots and leaves. Plants suffering from too dry soil conditions are most susceptible to infection. Therefore mulch to conserve moisture and water in dry periods before the soil dries out completely; this is particularly important for plants growing against walls. At the first sign of mildew spray with dinocap, benomyl, thiophanate-methyl or sulfur, or dust with sulfur. Up to four applications of any of these fungicides may be needed to keep the disease in check.

Downy mildew (*Plasmopara viticola*) occurs on vines quite frequently in the United States. It shows as light green blotches on the upper surface of the leaves and as a downy mildew on the lower surface of these patches. Affected areas dry up and become brittle causing the leaves to curl and fall. Diseased berries shrivel and become brown and leathery. The tips of the shoots may also be attacked. Cut out and destroy all diseased tendrils and leaves to remove the over-wintering stage of the fungus, although some spores also over-winter in the shoots and bud scales. Where this disease is known to be troublesome, spray before it develops using a protective fungicide such as zineb or maneb. Repeat the treatment at 10–14 day intervals, ceasing when further spraying would blemish the maturing fruit. Do not spray within seven days of harvesting the crops.

Nuts

COBNUTS AND FILBERTS
These trees are closely related and therefore susceptible to the same pests and diseases.

Shoots dying
Honey fungus (*Armillaria mellea*) can kill cobnut and filbert trees, and also walnut trees. The most important symptom of this disease is the shoots dying, the leaves of which wither but continue hanging for a while before falling. White fan-shaped masses of fungal growth develop beneath the bark of the roots and the main stems at or just above ground level. Dark brown root-like structures called rhizomorphs may form on the roots and grow through the soil to spread the disease. At the base of an infected tree honey-colored toadstools may develop in the fall. Dig out and burn dead and dying plants together with as many of the roots as possible. Sterilize the soil with a 2 per cent solution of formalin (made by dissolving 1 pt in 6 gal of water), applying 6 gal of this solution to each square yard of the infected area.

Leaves discolored
Faulty root action is caused by adverse soil conditions and can result in discoloration of the foliage. Consult the section on physiological disorders, pages 5–7.

Buds enlarged
Hazel big bud mites (*Phytoptus avellanae*) live inside the buds causing them to become swollen and rounded. The same symptom is produced by big bud mites on black currants but different species are responsible and there is no cross-infection between these plants. Infected buds usually fail to develop in spring but little harm seems to occur and control measures are not necessary.

Nuts maggoty
Hazelnut weevils (*Curculio neocorylus*) lay their eggs in the young nuts in May or June. The weevils' grubs feed on the kernels until July or August when they cut circular holes in the side of the nut-shells and drop to the soil where they pupate. If a large proportion of the crop has been damaged in previous seasons, spray with carbaryl at the end of May and again two to three weeks later.

WALNUTS
The troubles listed below may affect both the common walnut and the black walnut.

Shoots dying
Honey fungus (*Armillaria mellea*) can kill walnut trees. For symptoms and treatment, see under cobnuts and filberts.

Leaves blistered
Walnut blister mite (*Eriophyes erinea*) causes one or more raised areas to appear on the leaflets. The undersides of these raised areas are covered by fine pale cream hairs among which the mites live and feed. The blisters are oval and about $\frac{3}{4}$ in long. The mites cause no harm and can be tolerated.

Leaves discolored
Faulty root action is caused by adverse soil conditions and may result in the leaves being discolored. Consult the section on physiological disorders, pages 5–7.

Leaves and nuts spotted
Walnut leaf blotch (*Gnomonia leptostyla*, syn *Marssonina juglandis*) shows first as yellow-brown patches on the upper leaf surfaces, the undersurfaces of which become grey and then dark brown. The leaves wither and fall prematurely. Black or dark brown slightly sunken blotches also appear on the young green nuts. Rake up and burn fallen leaves and spray young trees in spring with zineb when the leaves are half-size, repeating three times at two week intervals.

Nuts abnormal
Walnut soft shell is a physiological disorder, the exact cause of which is not known. It is most troublesome after wet summers and on the fruit of trees growing on their own roots. The nuts develop a thin shell and, when dry, show holes at the apical end, thus letting in secondary organisms which soon rot any kernel that may have formed. Try to prevent this trouble by good drainage and cultivation. Alternatively grow only grafted trees.

ALMONDS
For pests and diseases of almonds, see references to ornamental *Prunus*, pages 83–90.

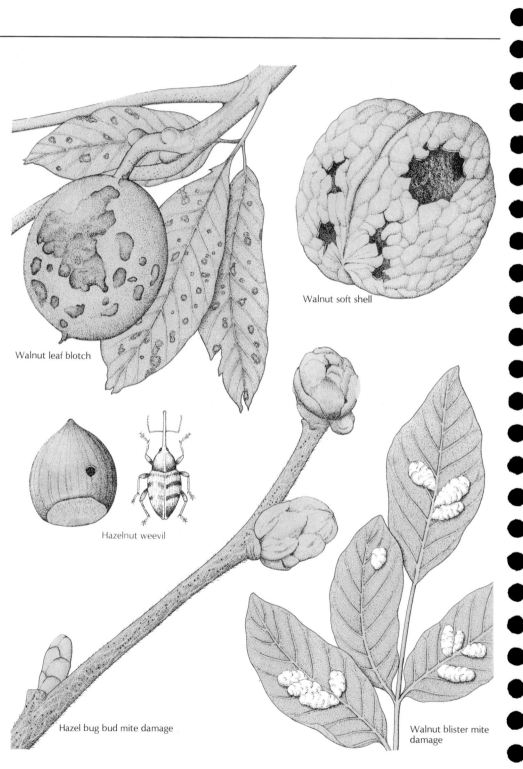

Walnut leaf blotch

Walnut soft shell

Hazelnut weevil

Hazel bug bud mite damage

Walnut blister mite damage

Figs

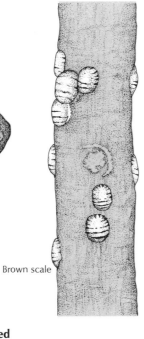

Gray mold

Fig mosaic

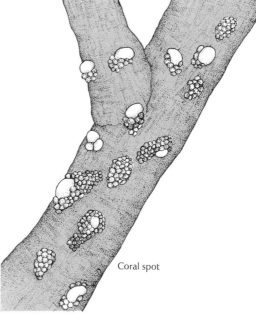

Brown scale
Coral spot

Fig canker

Leaves discolored

Faulty root action caused by adverse soil conditions can result in the leaves turning yellow or developing brown blotches. Whenever this occurs consult the section on physiological disorders, pages 5–7.

Red spider mite (*Tetranychus urticae*) is a tiny yellow-green creature that may damage figs growing in warm sheltered positions. The mites suck sap from the undersides of leaves, causing a yellow discoloration. When damage is seen the fig should be sprayed with malathion or a systemic insecticide three times at seven day intervals.

Leaves distorted

Fig mosaic virus causes the lobes of the leaves to become narrowed and irregularly shaped. Infected leaves may also become discolored in either of two ways. Yellow-green blotches up to $\frac{1}{2}$ in across with paler margins may develop, scattered at random over the leaf surface. Alternatively pale green spots or bands with narrow red-brown margins may form. These are associated with the larger veins. There is no cure for this disease and affected plants should be destroyed. It is

likely that most instances of the disease in the United States are due to propagation from an infected plant.

Stems with pests visible

Brown scale (*Parthenolecanium corni*) may encrust older branches. The scales consist of red-brown, convex shells about $\frac{1}{4}$ in long, beneath which the scale insects live and feed by sucking sap. Control them by spraying with a dormant oil in December or January. Alternatively spray the fig tree with malathion in early July and repeat two weeks later.

Shoots with visible fungal growth

Fig canker (*Phomopsis cinerescens*) appears on the bark as oval cankers which are sometimes large enough to kill whole branches. The diseased tissues are studded with minute pointed fruiting bodies of the fungus from which, in wet weather, extrude white tendrils of spores. These are spread by rain, birds and insects. Infection occurs through snags, pruning cuts and other wounds. If possible cut out and burn infected branches before the spores are produced. Disinfect pruning tools frequently during treatment. Cut off snags

flush with the branches and cover all wounds with a fungicidal wound paint.

Coral spot (*Nectria cinnabarina*) commonly causes die-back of shoots on which appear, towards the base of the dead wood, pink to red cushion-like pustules of spores. Treat as for fig canker (see above).

Fruits falling

Fruit drop is usually caused by an irregular supply of moisture at the roots. Prevent this trouble by mulching to conserve moisture and by watering in dry periods before the soil dries out completely.

Fruits rotting

Gray mold (*Botrytis cinerea*) can attack the fruit, causing it to fall prematurely or to rot and become covered with brown-gray velvety fungus growth. The fruits may also dry up and hang on the tree in a mummified condition. Remove and burn affected fruits otherwise the fungus may attack the young shoots through small wounds caused by frost damage and kill them for a distance of up to a foot or more. If this happens treat as for fig canker (see above).

Seeds and seedlings

Introduction
The problems dealt with in this section are those that commonly occur nowadays in gardens. Some pests and diseases that were troublesome in the past have been omitted because they hardly ever occur these days.

This is due partly to treatments carried out by seed merchants before the sale of seed or tubers, and partly to government restrictions on the varieties that can be sold which have, for example, helped to eradicate wart disease of potatoes.

Troubles that cause great losses only in commercial crops, such as virus diseases of lettuce, have also been omitted, as have those which occur so late in the season that they have very little effect on yield and quality even though producing severe symptoms, for example powdery mildews on peas and root crops.

In addition to the measures given here for the control of specific problems, the reader should also observe the basic measures listed in the Introduction on pages 2–3.

Seeds and seedlings

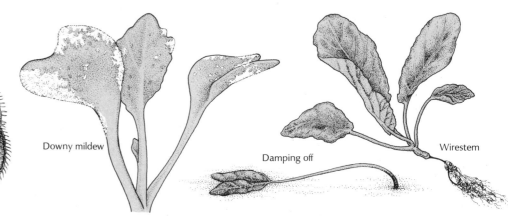

Seedcorn fly

maggot   damage

Millipedes

Woodlouse

Downy mildew

Damping off

Wirestem

Seeds eaten
Mice will dig up and eat larger seeds such as those of peas, beans and sweetcorn, autumn and spring sown crops being at greater risk than summer sowings. Holes can be seen in the earth where mice have tunneled down to find the seeds. If germination has begun, the shoot is usually not eaten and can be found discarded near the seed row. Repel mice by wetting the seeds with kerosene and dusting them with alum before sowing. Mousetraps can be set along the seed row, but they must be covered to stop birds and pets from setting them off.

Seedcorn maggot (*Hylemya platura*) can prevent the germination of all types of bean seed. The white, legless maggots grow up to $\frac{1}{5}$ in long and feed on the cotyledons and shoots. If the seedling does manage to grow, the shoot is often distorted and may be blind. Damage is more likely to occur in cold, wet seasons when germination is slow. Steps taken to hasten germination, such as using cloches to warm the soil and watering during dry spells, will enable the plants to pass more rapidly through this susceptible stage of growth. Treating the seed row with diazinon or chlorpyrifos granules will also give some protection against this pest.

Seedlings eaten
Slugs, woodlice and millipedes can destroy the seedlings of most vegetables by eating the stems and young leaves. Slugs can be controlled by scattering methiocarb or metaldehyde pellets along the seed rows. Methiocarb also gives some control of woodlice and millipedes. However, if these pests are numerous methiocarb will not give complete control, and they have to be tolerated.

Flea beetles (*Phyllotreta* spp) are tiny beetles, up to $\frac{1}{10}$ in long, which eat small holes in the upper surface of the seedling leaves of brassicas, turnips, swedes and radishes. As the young leaves grow they develop a very tattered and torn appearance due to the expansion of these holes. Several species of flea beetle may be responsible; they are generally black, sometimes with a yellow stripe on their wing-cases. Heavy infestations can check the growth of seedlings or even kill them, but plants are unlikely to suffer further damage once they have grown beyond this stage. Control the beetles by dusting the seedlings with either derris or pirimiphos-methyl. Keep the seed rows watered during dry weather until the plants are established and no longer susceptible to this pest.

Leaves discolored
Cold night temperatures can cause the foliage of seedlings to turn white, silver, yellow or purple. This is particularly true if there is a great difference between the day and night temperature. There is very little one can do to prevent such discoloration, but affected plants may benefit from applications of a foliar feed to help them overcome the check in growth.

Downy mildew, caused by the fungus *Peronospora parasitica*, affects young brassica seedlings, especially cauliflowers. The fungus *Bremia lactucae* causes downy mildew on lettuce. Both fungi show as a white mealy or furry growth on the undersides of the leaves, which become discolored. Affected seedlings may be severely checked by the disease. A moist atmosphere encourages the mildews and they are most likely to occur where the seedlings are growing too close together. Therefore sow seed thinly, on a fresh site each year in well drained soil with a good tilth. Thin out early. At the first signs of trouble remove affected leaves and spray with mancozeb or zineb, or use dichlofluanid on young cauliflowers, bordeaux mixture on other brassicas or thiram on lettuce seedlings, repeating as necessary.

Stems collapsing
Damping off causes the collapse of seedlings at ground level, especially of beetroot, celery, lettuce, peas and tomatoes. The symptoms usually appear soon after germination when the seedlings topple over due to the tissues at the base of each stem withering and blackening. The disease is only likely to be troublesome where seedlings are overcrowded or growing in wet conditions in compacted soil with poor aeration. Prevent damping off by sowing thinly in a good tilth and do not over-water. Use captan or thiram seed dressings to prevent this trouble. Check slight attacks by watering with captan, zineb or Cheshunt compound after removal of the dead seedlings.

Wirestem fungus (*Pellicularia filamentosa*, syn *Rhizoctonia solani*) attacks brassica seedlings, shrinking their stems at ground level and causing the roots to blacken and die. The seedlings eventually collapse. In slightly older plants toppling may not occur but the tissues at the bases of the stems remain hard, brown and shrunken, thus checking growth. Prevent this trouble by sowing thinly in a good tilth and avoiding over-watering. Control wirestem by raking quintozene into the soil before sowing.

Lettuce/Radish

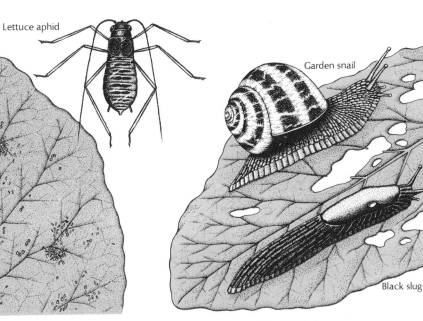

Lettuce aphid

Garden snail

Black slug

Cutworms

Flea beetle

LETTUCE

The greatest damage to lettuce is caused by pests, particularly aphids and cutworms.

Leaves with pests visible

Lettuce aphid (*Nasonovia ribisnigri*) is the greenfly most likely to attack lettuces. This aphid is green with black markings and may be found on lettuces from May to September. Heavy infestations can rapidly develop and cause the lettuces to become stunted and unpalatable. Control them by spraying with dimethoate, malathion or derris. Use derris if harvesting plants within a week.

Leaves eaten

Slugs and snails can cause serious damage in wet seasons and may kill young plants before they are established. These pests feed mainly at night so their presence may go unnoticed. They usually leave behind a silvery trail of slime on the foliage which helps to distinguish slug and snail damage from that caused by caterpillars. Protect lettuces by scattering slug pellets between the plants until the damage stops. Do not place pellets on the plants as they will rot and possibly allow gray mold to spread to the lettuce leaf.

Leaves blotched

Downy mildew (*Bremia lactucae*) can affect mature plants as well as seedlings, though it is usually only the older, outer leaves that show symptoms. Affected leaves may be attacked by secondary troubles such as gray mold, and rot fairly quickly. See facing page for symptoms and treatment.

Crowns rotting

Gray mold fungus (*Botrytis cinerea*) can infect the leaves but most commonly attacks plants at soil level, causing them to wilt. Prevent this trouble by providing the best soil conditions to encourage rooting, and plant out seedlings carefully to avoid damaging them. Use PCNB or dicloran raked into the soil before sowing or transplanting to help prevent gray mold.

Roots severed

Cutworms are soil-dwelling caterpillars of certain species of moth. They live near the soil surface and feed on roots and the bases of stems. The first sign of trouble is that the plant wilts and an examination of the roots will show that the main tap root has been partially or completely severed. A search of the soil near the dying plant will often reveal the cutworm. It is a dirty pale brown caterpillar up to $1\frac{3}{4}$ in long which tends to curl up when uncovered. Frequently there are only a few cutworms and they can be dealt with by hand. In new or neglected gardens there may be larger numbers since the moths lay more eggs in such ground. Treat the soil with chlorpyrifos or diazinon before sowing.

Roots with pests visible

Lettuce root aphid (*Pemphigus bursarius*) may attack plants sown between mid-April and late June. Infested plants can be recognized by their slow growth and tendency to wilt in sunny weather. If these plants are lifted the pale yellow aphids can be seen sucking sap from the roots. The aphids cover themselves and the roots with a white, mold-like, waxy substance. Once the roots are heavily infested there is little that can be done to save the plants, although drenching the roots with malathion at spray strength may check the aphids. The following year treat the soil with diazinon granules before sowing and grow lettuce on a different site. Alternatively, plant resistant varieties such as 'Avoncrisp' and 'Avondefiance'.

RADISH

Being quick to mature, radishes are generally free of diseases and are liable to attack by only two pests.

Leaves with holes

Flea beetles (*Phyllotreta* spp) cause small holes to appear in the foliage, especially while the plants are at the seedling stage. These beetles grow up to $\frac{1}{10}$ in long and are generally black, sometimes with a yellow stripe on their wing-cases. Keep the plants watered during dry spells to help them through the vulnerable seedling stage. If an insecticide is needed, apply derris dust as this has a short persistence and will not be present when the radishes are ready for eating.

Roots tunneled

Cabbage maggots (*Hylemya brassicae*) tunnel inside the swollen part of the root making a narrow brown tunnel. They also eat the seedling roots and check the plant's growth. Attacks are likely to occur on sowings made between mid-April and August. In areas where this pest is troublesome protect sowings by treating the seed row with diazinon or chlorpyrifos.

Brassicas 1

This section covers the following brassicas: cabbage, cauliflowers, Brussels sprouts, broccoli, and kale. Swedes and turnips, which are also brassicas, are discussed on page 47.

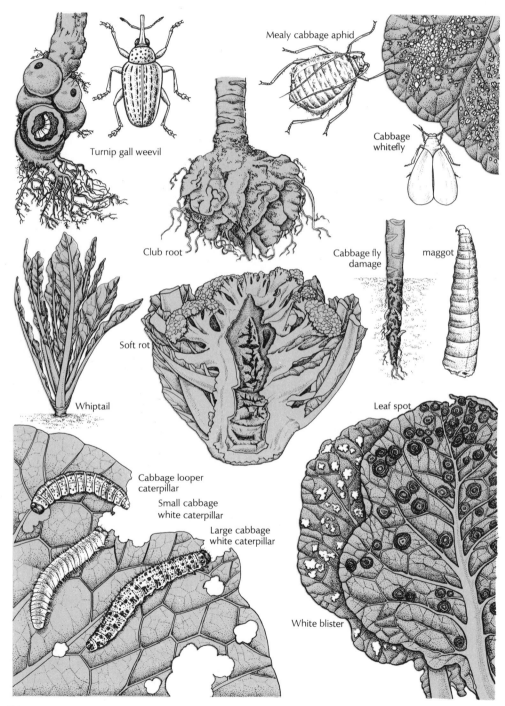

Turnip gall weevil

Mealy cabbage aphid

Cabbage whitefly

Club root

Cabbage fly damage

maggot

Soft rot

Whiptail

Leaf spot

Cabbage looper caterpillar

Small cabbage white caterpillar

Large cabbage white caterpillar

White blister

Leaves with pests visible

Cabbage whitefly (*Aleyrodes proletella*) is a tiny, white, moth-like insect found on brassicas throughout the year. Both the adults and their flat, oval, scale-like larvae feed by sucking sap from the undersides of the leaves. The adults fly up from plants when they are disturbed. Heavy infestations can lead to the foliage becoming covered with a superficial black mold that grows on the whiteflies' sticky excretions; apart from this, little real damage is caused. The pest mainly infests outer leaves rather than the edible parts of the plant. Cabbage whitefly is different from the whitefly species that occurs in greenhouses, and it attacks only brassicas. Light infestations can be tolerated, but if whiteflies become too numerous they can be controlled by spraying thoroughly with a pyrethroid compound or pirimiphos-methyl. The immature stages are less susceptible to the insecticide than are the adults, so three or four applications at seven day intervals may be necessary.

Mealy cabbage aphid (*Brevicoryne brassicae*) occurs in dense colonies on the undersides of brassica leaves between late spring and the fall. These blue-gray aphids feed by sucking sap, and turn the foliage yellow. Young plants are particularly vulnerable, and aphid attacks on the new leaves and the growing point can cause severe stunting. Infestations can also occur on Brussels sprout buttons and in the hearts of cauliflowers and cabbages, making them inedible. The aphid over-winters as eggs laid on the stems of brassicas. Burn old brassica plants or bury them deeply as soon as cropping has finished. This prevents eggs hatching in the spring and spreading aphids to new plants. Inspect all brassicas regularly during the summer. Spray with dimethoate, formothion or menazon if aphids are present. Plants that are being transplanted can be treated by dipping the foliage into a tin of spray-strength insecticide. Wear rubber gloves when doing this.

Leaves with holes

Cabbage caterpillars include the three most common caterpillars that feed on brassicas; those of the large cabbage white butterfly (*Pieris brassicae*), the small cabbage white butterfly (*Pieris rapae*) and the cabbage looper (*Trichoplusia ni*). All have two generations during the summer and damage to plants occurs between April and October. Caterpillars of the large cabbage white are yellow with black markings, whereas those of the small cabbage white are pale green with a velvety appearance. Cabbage looper caterpillars are green or brown and are without any obvious covering of hairs. The caterpillars eat irregular-shaped holes in the foliage and often burrow into the hearts of cabbages. Control measures should be aimed at the young caterpillars before they penetrate the heart leaves, because they are difficult to get at once they do so. Regular inspections during the summer will reveal when damage is starting. Removing the caterpillars by hand is effective for a small number of plants; otherwise dust the plants with carbaryl, or spray with either fenitrothion, trichlorphon or *Bacillus thuringiensis*.

Wood pigeons (*Columba palumbus*) peck at the foliage until the leaves are reduced to only the midribs and larger veins. Pieces of torn foliage may be left scattered on the ground. Damage can occur throughout the year, but brassicas are most at risk in winter. Scaring devices and bird repellant sprays are of limited value in deterring pigeons. In areas where they are a persistent problem, some form of vegetable cage or netting is required.

Leaves distorted

Whiptail (molybdenum deficiency) is a disorder of broccoli and cauliflowers which reduces the leaf-blades to the midrib so that they become thin, strap-like and ruffled. The heads of affected plants fail to develop. Whiptail is caused by a lack of molybdenum, a trace element that is either lacking or unavailable to plants in some acid soils. It can be prevented to a certain extent by liming before sowing or planting. Apply a solution of sodium molybdate at the rate of 1 oz in $2\frac{1}{2}$ gal of water to each 10 sq yd of soil.

Leaves spotted

Leaf spot fungi include *Mycosphaerella brassicicola*, *Alternaria brassicae*, *Alternaria brassicicola*, the most common in gardens being *Mycosphaerella brassicicola*. It usually affects

Brassicas 2

the older leaves, causing round brown spots. These show concentric rings of pinpoint-sized bodies which produce the spores of the fungus. Sometimes affected tissues fall away, leaving holes. Leaf spot is most troublesome in wet seasons, particularly on plants that grow too soft because of heavy dressings of nitrogenous fertilizers. Remove and burn affected leaves. Where plants are growing close together, remove alternate plants to improve the circulation of air around them, since this will help to prevent further trouble. The following season, grow brassica plants in a different part of the garden.

White blister (*Albugo candida*) causes unsightly glistening white masses of fungus spores on the leaves of affected plants. The masses show either singly or in concentric rings on the leaves, particularly those of Brussels sprouts. It can also distort the heads of cauliflower and broccoli, but apart from this does little damage; affected plants are edible and there is no reduction in yield. The most severe attacks occur on overcrowded plants so dig up alternate plants after removing and burning affected leaves. Rotation of crops also helps to prevent reinfection of brassicas the following year.

Leaves rotting
Gray mold and soft rot are often found on rotting leaves and stems, following frost damage or some other type of injury in a wet season. The gray mold fungus (*Botrytis cinerea*) produces a gray-brown growth on the affected parts and bacterial soft rot (*Erwinia carotovora*) causes the parts to turn slimy and produce an unpleasant odor. Rotting caused by either organism is particularly common in the centers of young white cabbages that have grown too soft. Prevent soft growth by the restrained use of nitrogenous fertilizers and by applying a dressing of sulfate of potash before planting. Remove and burn severely affected plants to stop these diseases from spreading. When frost is forecast lift mature heads to prevent frost damage and subsequent infection.

Stems galled
Hormone weedkiller damage is caused by the misuse of hormone weedkillers. It can

severely damage cabbages and Brussels sprouts, and affected plants have rough wart-like out-growths on the stems, especially at the base. These should not be confused with the symptoms of club root. The leaves may become somewhat narrowed and the plants remain stunted and fail to produce an edible crop. Plants showing such severe symptoms should be destroyed. Other brassicas may show slight symptoms but usually produce an edible crop and so can be retained. Damage of this type can be prevented by the correct use of weedkillers, which should be applied only on a still day using apparatus that is kept solely for the application of weedkillers and not for general watering or spraying.

Roots eaten
Cabbage maggots (*Hylemya brassicae*) eat the fine roots until all that remains of the root system is a blackened rotting stump. Several generations of maggots occur between early May and September. Seedlings and recently transplanted brassicas are the most susceptible; well established plants can often tolerate attacks without any apparent ill effects. Damaged plants stop growing, tend to wilt during sunny weather, and the foliage turns an unhealthy, blue-green color. Young plants are frequently killed. Treat seed rows and the soil around transplants with soil insecticides such as diazinon, chlorpyrifos, bromophos or azinphos-methyl. Check attacks on established plants by applying a spray-strength solution of trichlorphon to the soil, drenching it thoroughly.

Roots galled
Club root (*Plasmodiophora brassicae*) causes plants to become stunted, particularly if they are infected at the seedling stage, although older plants may merely show discolored leaves which wilt on hot days. These symptoms occur because the roots are unable to function normally, having become distorted into a thickened and swollen mass due to infection by the soil-borne organism. New infections are usually introduced into gardens on seedlings. Check that the roots of bought brassicas and related plants are healthy. Club root may develop after severely diseased seedlings contaminate a badly drained alka-

line soil. However the organism flourishes best on very acid soils, in which it can be controlled to a certain extent by applying a dressing of ground chalk or limestone at 14 oz per square yard. In succeeding years apply lime at a rate of 8 oz per square yard. On heavy soils improve the drainage by thorough cultivation, deep digging and the incorporation of plenty of humus. Destroy all cruciferous weeds as they can act as hosts for the organism. Chemical control of club root can be achieved by applying PCNB to the seedbed before sowing brassicas. Spraying young plants with this pesticide when transplanting them into their final positions also helps to prevent infection.

Turnip gall weevil (*Ceutorhynchus pleurostigma*) is a beetle which causes damage that is often confused with club root disease because both result in swelling of the roots. Galls caused by the weevil are roughly rounded; when cut in half, they reveal a small white grub in the hollow center. If the grub has already left the gall to pupate in the soil there will be a small round hole in the side of the gall. This is not a damaging pest on established brassicas because the galls do not seem to interfere with root action and growth remains unaffected. However, seedlings may have their growth checked. Help to prevent this by sowing at the correct depth and by thinning out seedlings early.

HEARTLESS AND SPLITHEADED CABBAGES

Split head

Lack of heart

These troubles are caused mainly by faulty root action. Failure to heart can occur if there is not enough humus in the soil or if the soil has not been firmed sufficiently before and after planting. This disorder is likely to be worse on a shady site, and it is also encouraged by drought. Split cabbages are usually caused by long periods of dry soil conditions followed by irrigation or heavy rain. It is advisable therefore to water before the soil dries out completely as well as to make sure that cabbage seedlings are planted carefully in well prepared soil. Frost damage can also cause split heads (see Leaves rotting).

Asparagus/Celery and celeriac 1

Asparagus

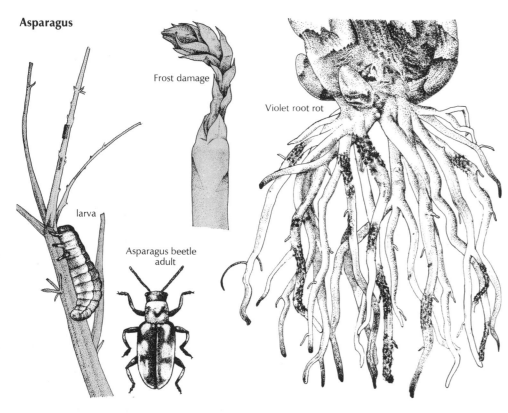

Frost damage

larva

Asparagus beetle adult

Violet root rot

Celery and celeriac

Celery leaf miner maggot

Leaf spot

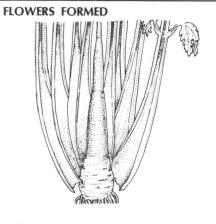

FLOWERS FORMED

Celery should not produce flowers but plants frequently bolt (go to seed) when the soil is too dry at a critical stage in their growth. However, bolting is often not noticed until lifting time when the heart of the plant is found to consist of just one round fat inedible flower stalk. Prevent this trouble by watering before the soil dries out completely, particularly when the seedlings are still in the seed box and just after planting out.

Leaves and stems eaten

Asparagus beetles (*Crioceris asparagi*) first appear on the plants in May or June when they lay small black eggs on the foliage. The beetles are about $\frac{1}{4}$ in long and have black wing-cases with square yellow markings and red margins. The larvae are gray-black and, like the adults, feed on the foliage and stems. The plants may become badly defoliated and the stems can be girdled by the beetles chewing away the surface tissues. This causes the stem above the point of damage to dry up and turn yellow-brown. Spray with derris, malathion or pirimiphos-methyl. These insecticides will control both the adults and larvae. Carbaryl will control adults only.

Stems blackened and/or distorted

Frost damage is common in late spring. It causes edible shoots to blacken and die. Less severe damage causes twisting or bending of stems near the tips or slight pinching

at soil level with a consequent withering of the tips. Cut off shoots showing any of these symptoms as they will not recover.

Roots dying

Soil-borne fungi can attack and kill both the crowns and roots of asparagus plants. The fungus responsible may be either violet root rot (*Helicobasidium purpureum*) or fusarium wilt (*Fusarium oxysporum*). With the death of the roots the top growth turns yellow and dies, and a gap develops in the asparagus bed as the disease gradually spreads outwards. Where only slight infection has occurred, isolate the diseased area by sinking sheets of thick polyethylene into the soil to a depth of 1ft. In severe cases scrap the infected bed and make a new one on a fresh site. Keep the diseased area free of asparagus for several years; avoid also root crops and control susceptible weeds such as dock, bindweed and dandelion.

Leaves mined

Celery leaf miner maggots (*Euleia heraclei*) live inside the leaves of celery and related plants such as parsnip and lovage. Attacks begin from late April to June and there is a second generation from July to September. The second generation tends to be more numerous and, by the end of the summer, most of the foliage may be destroyed. The small white maggots cause irregular pale green blotches to appear on the leaves. Later the tunneled areas dry up and turn brown, giving the plants the appearance of being scorched by fire. Although less numerous, the first generation maggots can be more damaging since they can check the growth of young plants, causing the celery sticks to become stringy and bitter. Control light infestations by picking off and burning the affected parts of the leaves. Spray heavier attacks with either of the insecticides trichlorphon or dimethoate.

Celery and celeriac 2

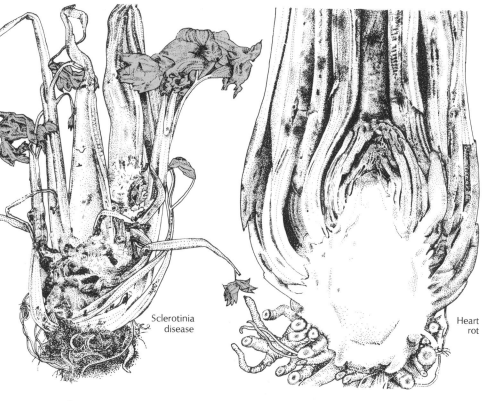

Sclerotinia disease

Heart rot

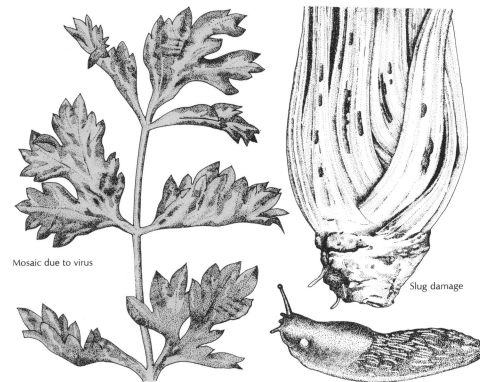

Mosaic due to virus

Slug damage

Leaves spotted

Leaf spot is caused by the fungus *Septoria apiicola* and appears as small brown spots on which develop pinpoint-sized black fruiting bodies of the fungus. The stalks may also be affected and in severe cases yield is greatly reduced. However, as most reliable seed firms treat their seed against this disease it is only likely to be troublesome in some varieties of self-blanching celery. As soon as any symptoms are seen spray with a copper fungicide, mancozeb, maneb, chlorothalonil or zineb, repeating several times at two week intervals.

Leaves discolored

Viruses including arabis mosaic and cucumber mosaic viruses can cause a yellow and green mottling of the leaves which may also be distorted. Affected plants are usually stunted and should be destroyed immediately symptoms are seen.

Stems eaten

Slugs may cause considerable damage to celery as they crawl between the stems, causing them to become brown and pitted where the slugs have rasped away the surface tissues. Secondary rots can enter through these wounds and ruin the plants. Kill slugs by scattering slug pellets containing methiocarb among the plants, particularly during spells of warm weather after heavy rain—these conditions encourage slugs to come to the soil surface when they are more likely to find and feed on the pellets.

Stems cracked

Faulty root action due to dryness at the roots can cause lengthways splits in celery stalks. Similar symptoms can also occur on plants grown too soft due to excess nitrogen in the soil. Prevent this trouble by digging in plenty of humus and mulching well around the plants to conserve moisture, and by watering

before the soil dries out completely. Be sparing with nitrogenous fertilizers, or harden growth with sulfate of potash.

Boron deficiency causes brown horizontal cracks across the stalks and leads to poor growth with the leaves turning yellow and dying. Before planting in soil where this trouble is known to occur apply borax (sodium tetraborate) at the rate of 1oz to 20 sq yd. Mix the borax well with a large amount of light sand to facilitate even spreading.

Stems rotting

Heart rot is caused by the soft rotting bacterium *Erwinia carotovora*. The disease is not usually noticed until an affected plant is lifted when it shows a brown slimy mass in the center which often extends up the stalks making the plant worthless. Even if the trouble is noticed earlier a diseased plant cannot be saved and should be burned. The bacteria

can only enter celery through wounds, so control slugs with methiocarb pellets (see under Stems eaten) and earth up plants carefully to avoid damaging them. Change the site for celery every few years to avoid a build-up of bacteria in the soil. Give some form of protection to plants when frost is likely.

Sclerotinia disease is caused by the fungus *Sclerotinia sclerotiorum* and shows as a rapid rotting of the stems. White fluffy fungal growth develops on the diseased tissues and soon forms hard black resting bodies which are $\frac{1}{2}$ in or more in length. These structures fall into the soil and remain dormant there during the winter, germinating the following spring to cause new infections. Destroy any plant showing such symptoms, preferably before the resting bodies form. Help to prevent reinfection the following year by growing celery on a fresh site and by dusting seed with benomyl before sowing.

Peas and beans 1

Leaves eaten

Pea and bean weevil (*Sitona lineatus*) is a gray-brown beetle about $\frac{1}{4}$ in long which often escapes notice since it drops to the ground if the plant is disturbed. The adults feed by eating U-shaped notches from the leaf margins of peas and broad beans. Other beans are not attacked. Damage is frequently seen during the early summer but the effects on the plants are slight. Control measures are only necessary if seedling plants are being attacked, in which case apply pirimiphos-methyl dust to the plants.

Leaves spotted

Chocolate spot is a disease of broad beans caused by the fungus *Botrytis fabae*. Small chocolate-colored spots or streaks develop on the leaves and stems, and may coalesce until large areas are discolored. In a wet spring, affected plants may then become blackened and die, especially winter-sown crops. Plants grown in too acid a soil, or which have become soft through excessive nitrogenous fertilizers, are more susceptible to attack. Therefore sow seed thinly in well drained soil, adding sulfate of potash at $\frac{1}{2}$ oz per square yard before the November sowing. Lime the plot to give a pH of 7, the optimum value for broad beans. Burn any diseased plants at the end of the season to prevent over-wintering of the fungus. The following year, spray with a copper fungicide as soon as the foliage appears.

Halo blight (*Pseudomonas phaseolicola*) is a seed-borne bacterial disease of dwarf, French and runner beans. An infected plant will show angular spots on the leaves surrounded by a lighter-colored halo. Later, greasy-looking lesions develop on the stems and pods. These spots turn red-brown but in wet weather a white encrustation shows on them as bacteria ooze out. Prevent the disease by buying only good quality seed. Should the disease appear nothing can be done and all plants are likely to be infected, so wait until the end of the season before burning them.

Viruses such as pea and bean mosaic viruses cause stunting of plants which bear leaves mottled with dark and pale green or green and yellow areas. The pods may also be mottled and distorted and, in the case of peas, also rough and ridged. Dead patches or streaks may also show on the leaves of pea plants, which may also die back. Burn all infected plants and wash hands and tools after plants have been handled to prevent the virus from spreading.

Leaves and stems with pests visible

Greenfly and blackfly (*Acyrthosiphon pisum* and *Aphis fabae*) commonly attack peas and French, runner and broad beans, respectively. Both types of aphid are likely to be found on the shoot tips and undersides of the leaves during the summer months. Heavy infestations reduce the yield of peas and beans. Prompt treatment with an insecticide such as dimethoate, formothion, menazon, diazinon, demeton or pirimicarb will prevent damage if applied when aphids are first seen. Beans rely on bees for pollination, so if it is necessary to spray while they are in flower apply pirimicarb at dusk as this insecticide is relatively harmless to bees.

Stems wilting

Fusarium wilt (*Fusarium solani* f *phaseoli*) affects dwarf and runner beans, and causes the plants to wilt. The fungus is soil-borne but can sometimes be seen as a pink layer of fungal growth on red-brown sunken lesions at- the base of wilting stems which bear yellowing leaves. Red-brown longitudinal streaks will be seen in the internal tissues if an affected stem is cut open near ground level. Burn affected plants and avoid this disease by rotating crops so that beans are not grown on the same site in successive years. Grow only those varieties that are resistant to the disease.

Foot and root rot can be due to any of the fungi *Aphanomyces euteiches*, *Thielaviopsis basicola*, and species of *Phytophthora*, *Pythium* and *Fusarium*. These fungi, which are soil-borne, may attack both peas and beans. They kill the roots and often rot the stem-bases, turning them brown or black. The foliage becomes discolored and the plants may collapse or fail to produce good pods. Severely diseased plants should be destroyed but it may be possible to save less affected plants by watering them with a solution of

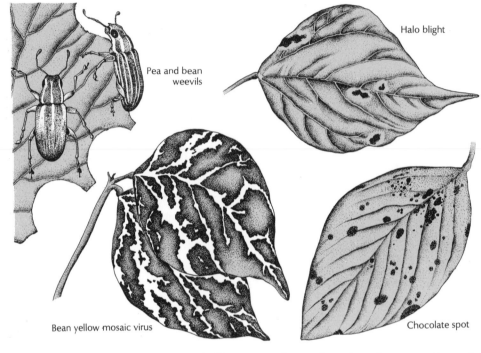

Pea and bean weevils

Halo blight

Bean yellow mosaic virus

Chocolate spot

captan, Cheshunt compound or zineb. These fungi are only likely to be troublesome where leguminous plants are grown on the same site for successive years. Therefore rotate crops to prevent the trouble. Use of a seed dressing containing captan or thiram may also help to prevent infection.

Pea early browning and bean yellow mosaic viruses can cause the collapse of pea plants. Affected plants usually show characteristic brown dead patches or streaks on the stems and leaves, or a yellow mottling of the foliage, before they collapse. Burn diseased plants and spray with appropriate insecticides regularly to control aphids since these viruses are transmitted by aphids as they feed.

Pods discolored

Pea thrips (*Kakothrips robustus*) attack the pods of peas. They are tiny thin insects up to $\frac{1}{10}$ in long which suck sap from the foliage and pods. Thrips are black when fully grown but are yellow in their nymphal stages. Their feeding causes a silvery-brown discoloration on the pods which may be distorted and contain only a few peas. Heavy infestations are most likely to occur in hot dry summers. Control thrips by spraying thoroughly with fenitrothion or malathion as soon as signs of damage are noticed.

Pods spotted or discolored

Pea pod spot (*Ascochyta pisi*) is a seed-borne disease. A plant raised from an infected seed may show dark brown sunken patches on the leaves and stems but the most serious trouble is due to the development of such spots on the pods; the damage can be severe in a wet season. Fruiting bodies of the fungus form on the spots and produce very many spores which spread the disease to adjacent healthy plants. The fungus grows through the pod to attack the seeds, which show brown or purple spots. Do not save seed from a diseased crop; lift and burn affected plants as they appear. As infection can also arise from the diseased debris of a previous crop, it is advisable to burn all haulms of a badly diseased crop after harvest and to refrain from growing peas on the infected site for several

Peas and beans 2

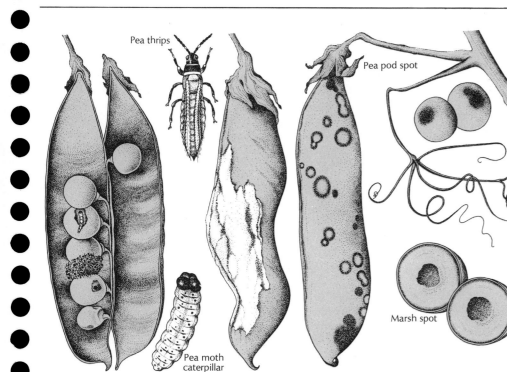

Pea thrips

Pea pod spot

Marsh spot

Pea moth
caterpillar

years. Seed bought from a reputable firm should be sown on a fresh site each year. Treat seed with a dressing containing captan or dry thiram if, because of a shortage of space, infected ground must be used again since this may enable a crop to be taken.

Anthracnose of dwarf beans is caused by the fungus *Colletotrichum lindemuthianum*, and it shows as black or red-brown areas on the pods and less conspicuously on the stems and leaves. The disease can be severe in cold wet weather and on plants being forced under glass, when a thin white crust may develop on the spots. It may be possible to check the disease if the plants are sprayed with maneb or zineb, repeating at two week intervals until flowering time. Burn all the plants at the end of the season. The fungus passes from the pod into the seeds causing brown or black patches on the skin; the disease is thus seed-borne and no seed should be taken from a diseased crop. Before sowing, dust seed with captan or thiram, and sow on a fresh site each year since the disease may survive in the soil on decaying debris.

Pods rotting
Gray mold fungus (*Botrytis cinerea*) occasionally rots the pods of leguminous plants in a wet season. The rotting tissues are covered with a gray-brown furry growth. Space the plants well to allow a good circulation of air since gray mold is encouraged by growing plants too close together. Pick off and burn diseased pods and protect remaining pods by spraying with benomyl, captan or thiophanate-methyl. Should the trouble occur every year, spray at flowering time.

Peas maggoty
Pea moth (*Laspeyresia nigricana*) lays its eggs between June and mid-August on flowering pea plants. The caterpillars make their way into the pods where they feed on the developing seeds. Pods picked late in the summer frequently have a large proportion that contain pea moth caterpillars. These larvae are about $\frac{1}{4}$ in long and are a creamy white color with black heads. Quick maturing varieties of pea that are sown early or late and flower outside the moth's egg-laying

period escape damage. Control measures are only necessary on those peas that come into flower within this period. Spray such plants thoroughly at dusk with fenitrothion seven days after the start of flowering.

Peas discolored internally
Marsh spot is caused by a lack of manganese, and shows as a dark rust-red spot or cavity in the center of peas. The pods look normal but the foliage of affected plants may show slight yellowing between the veins. If this is a recurrent problem spray with 2 oz of manganese sulfate to 3 gal of water plus a spreader, two or three times at two week intervals. Alternatively rake either fritted trace elements or a chelated compound into the soil before sowing.

Roots with pests visible
Bean root aphid (*Smynthurodes betae*) infests the roots of runner, French and sometimes broad beans during the summer. This aphid is a creamy-brown color and can be difficult to detect among the soil and roots; but its presence is indicated by the white waxy powder that the aphids secrete. This white dusting of the insects and roots can be mistaken for a mold. The aphids suck sap from the roots, and this can lead to poor growth, wilting and reduced cropping. In gardens where root aphid is a regular problem treat the seed drill at sowing time with diazinon granules. Treat established plants by drenching the soil with spray-strength malathion, although if the plants have already started to wilt it is probably too late for effective treatment. Do not grow beans in the same soil in consecutive years if root aphid is known to have been present.

Roots rotting
Soil-borne organisms may rot the roots of leguminous crops, although the trouble is only likely to occur where no rotation of crops is carried out. The most obvious symptoms caused by root rot are a general discoloration of the foliage, failure to set pods and sometimes the complete collapse of the plants. Less affected plants may be saved by watering with either captan, zineb or Cheshunt compound.

FAILURE TO SET

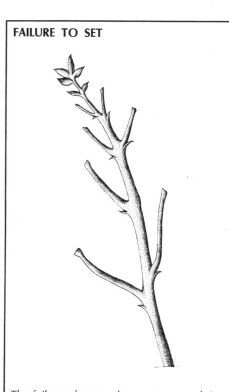

The failure of runner beans to set pods is a widespread problem and can be due to a number of causes. The most common is insufficient water but in some cases it is due to bird damage and, in some cold or wet seasons, may be caused by a lack of pollinating insects. The maximum amount of insect activity will be encouraged if runner beans are sown in a sheltered area and in a block of several short rows close together rather than one long row across the garden. Prevent dryness at the roots by watering throughout the flowering and fruiting period, applying 1 gal per square yard every two or three days. Harvest suitable pods regularly to prevent suppression of further pod set. Where crops have failed before, try growing white- or pink-flowered varieties which are less susceptible to this trouble than the red-flowered types. However, if the trouble is found to be due to birds, netting is the only method of prevention.

This section covers not only onions and leeks, but also the related vegetables garlic, chives and shallots.

Onions and leeks 1

Leaves discolored

Onion thrips (*Thrips tabaci*) are tiny, thin insects up to $\frac{1}{10}$ in long which feed by taking sap from the leaves. Thrips are black when fully grown but yellow during their nymphal stages. Their feeding causes a silvery-white discoloration of the leaves. Heavy infestations may occur in hot dry summers, and the growth of young plants may be checked. Light infestations can be tolerated but if leaves are being extensively marked, spray with malathion or azinphos-methyl.

Leaves withering

Downy mildew (*Peronospora destructor*) causes onion leaves to turn gray then wither and fall over. The disease is most troublesome in wet weather when an unobtrusive purple coating of fungus may appear on the diseased tissues. At the first signs of trouble spray with mancozeb or zineb, or dust with powders of either chemical when the plants are moist. In gardens where the disease is known to occur spray before any symptoms are seen. Grow onions on a fresh site each year in a well drained soil and keep weeds down to avoid a stagnant atmosphere. Do not store bulbs of affected onions; usually they are found to be soft at lifting time.

Leaves with obvious fungal growth

Rust (*Puccinia allii*) is most common on leeks but it can also affect chives and occasionally onions, shallots and garlic. It shows on the leaves as elongated pustules of bright orange spores. There is no control for rust but diseased plants usually produce good quality edible parts. Burn infected debris and raise leeks on a fresh site each season in soil which has not received much nitrogen. Alternatively rake in $\frac{1}{2}$ oz of sulfate of potash per square yard before planting or sowing.

Leaves and stems distorted

Onion eelworms (*Ditylenchus dipsaci*) are tiny worm-like creatures that live inside the tissues of the leaves and bulb. They are too small to be seen with the naked eye but their presence is readily detected by the symptoms they cause, especially on young plants. Plants that are attacked during the early stages of their growth may become

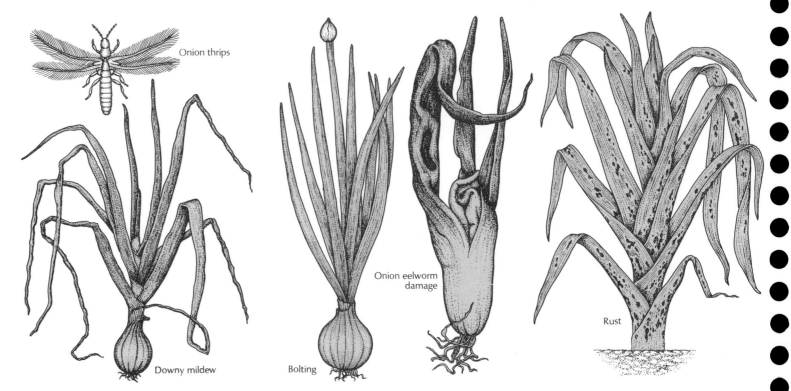

Onion thrips

Downy mildew

Bolting

Onion eelworm damage

Rust

abnormally swollen, twisted and distorted. This condition is sometimes known as onion bloat. The leaves and stem have a loose, mealy texture and lack their normal crispness. Such plants invariably rot and die before they reach maturity. Plants that do not become infected until they are fully grown will produce bulbs but they will have the characteristic mealy softness and are often split open at the base. Do not store infested bulbs since they frequently rot and, moreover, eelworms continue breeding in stored bulbs and will spread to healthy ones. There are no chemicals available to amateur gardeners for controlling eelworms, therefore dig up and burn infested plants. Onion eelworm can also attack other vegetables, including beans, carrots, parsnips and some weeds. These other host plants do not suffer severe damage in gardens but they can allow the pest to persist in the absence of plants of the onion family. Two types of vegetables which are not attacked and can be grown in infested soil

are lettuce and brassicas. Where eelworms have occurred keep the ground free of host plants, including weeds. It should then be possible to replant with onions and similar plants after two full years.

Flowers produced

Bolting may occur if plants are affected by drought at a critical stage of growth. It is essential, therefore, to ensure that the soil never dries out completely. Other causes are sowing too early, insufficient firming or planting during cold weather. If they appear, the flower stalks should be cut off immediately, and the bulbs can then be lifted at the normal time. They should be used first as they may not keep if placed in storage.

Roots and bulbs eaten

Onion fly maggots (*Hylemya antiqua*) occur between June and August. All members of the onion family can be attacked but onions themselves are the most susceptible. Young

plants can be killed by the maggots destroying the roots, and older plants can be made useless by the maggots tunneling into the base of the bulb and allowing rots to enter. The white maggots are $\frac{3}{8}$ in long when fully grown and these pupate in the soil. There are sometimes two generations during the summer but most of the pupae from the first generation over-winter without any further development that year. In gardens where onion fly is a problem, protect sowings or plantings made in May and June during their vulnerable early stages by treating the soil with diazinon or chlorpyrifos granules.

Bulbs rotting

White rot is caused by the fungus *Sclerotium cepivorum* and shows as a white fluffy growth on the diseased roots and tissues at the base of the bulb. Yellowing and subsequent death of the leaves occur. Burn affected plants before the small black resting bodies of the fungus can develop on the

Onions and leeks 2

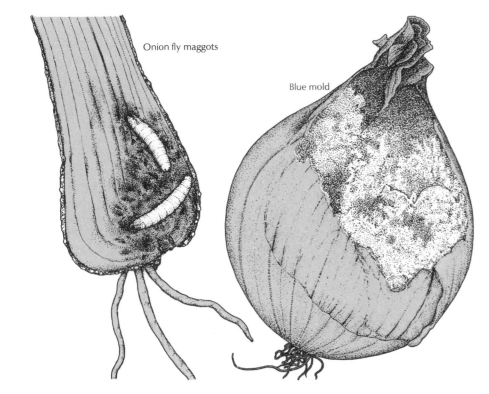

Onion fly maggots

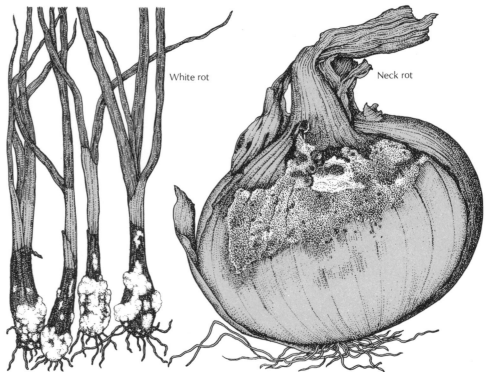

Blue mold

White rot

Neck rot

diseased tissues; these soon fall off into the soil and remain there where they overwinter. The following spring they are stimulated to germinate by the roots of newly developing onions and related plants. Such crops should, therefore, be grown on a different site each year. Treat the soil with Vapam three weeks before planting and while soil temperatures are above 13°C/55°F, or dust the open furrow with benomyl before and after sowing. Spray with double strength benomyl when plants are 7–8 in high.

Neck rot fungus (*Botrytis allii*) can cause considerable loss of stored onions. A gray moldy growth develops on or near the neck of an affected bulb, which subsequently becomes soft and rotten. Later, large black resting bodies of the fungus develop on the rotting tissues. Grow onions well to produce hard, well ripened bulbs and do not apply fertilizers late in the season. Dry them after lifting and store only hard bulbs in a cool, dry place where there is free circulation of air

around them. Examine bulbs frequently and remove rotting onions as they are seen. Buy good quality seed which has been treated against neck rot, or sets from a reputable grower, and dust seeds and sets with dry benomyl before sowing or planting.

Blue mold and soft rotting bacteria (*Penicillium* spp and *Erwinia carotovora*) are both capable of causing storage rots. Blue mold causes onions to become soft and develop blue-green growths of fungus between the outermost scales. Soft rotting bacteria cause them to become soft and slimy with an offensive smell. These troubles occur as a result of incorrect harvesting or poor storage. Aim to achieve hard, well ripened bulbs by good cultivation, then harvest them properly and dry off quickly, twisting off the tops when they have dried. Store in a cool place on wire netting or slatted shelves, or hang them in strings or net bags from rafters. Examine onions regularly and remove any that are rotting or sprouting.

FAULTY ROOT ACTION

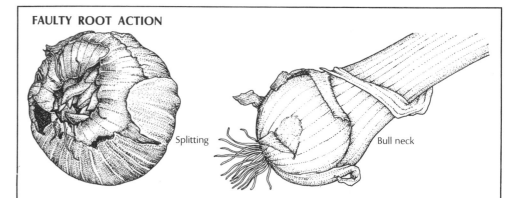

Splitting

Bull neck

This can result in a number of different troubles. In addition to bolting, a prolonged period of drought, followed by heavy rain or watering, can cause splitting of onions at the base. Drought can also cause an onion set to split into two separate bulbs. An excess of manure or too much nitrogen may result in the development of onions with abnormally thick necks (bull neck), but this trouble may also occur in wet seasons. Correct cultivation is therefore essential. Do not store any bulb showing any of these symptoms as it will not keep.

Potatoes 1

Leaves with pests visible

Aphids or greenfly may infest the stems and leaves of potatoes during the summer. Some types can spread virus diseases via their mouthparts, but this is not a serious problem unless some of the tubers are going to be saved for next year's seed potatoes. Heavy infestations of aphids may cause some leaf curling but otherwise their effects on the plants are not too damaging. If aphids do become numerous during the early part of the summer it may be worth while controlling them by spraying with pirimicarb, malathion or a systemic insecticide.

Colorado beetle (*Leptinotarsa decemlineata*) is a serious defoliator of potato and related plants. Originally native to western North America, it is now found throughout the United States and Canada. The beetles are about $\frac{1}{2}$ in long and are yellow with black stripes running lengthways along the wingcases. The larvae also grow up to $\frac{1}{2}$ in long and are red with two rows of black markings along their sides. Both adults and larvae feed on potato foliage and large numbers can rapidly reduce plants to bare stalks. Colorado beetle (which is also known as potato beetle) is one of the most serious pests of white potato in the United States. These beetles can be controlled by applying any of the insecticides carbaryl, azinphos-methyl or endosulfon. In some States, the new synthetic pyrethroid-type insecticides are available. All these chemicals give good control over both adult and immature Colorado beetles. Whichever insecticide is used, it is important to apply it as soon as symptoms of damage are noticed.

Leaves dying

Potato cyst eelworms (*Globodera rostochiensis* and *G. pallida*) live inside the roots but the symptoms are seen first on the foliage. The leaves on the lower part of the stems turn yellow, die prematurely and hang down, leaving a tuft of green leaves at the top of the stems. The plants are often killed well before the end of the summer and consequently the tubers are undersized and may be no larger than $\frac{1}{2}$ in across. A careful examination of the roots will reveal the mature female eelworms, they are spherical, pinhead-sized objects

known as cysts and may be white, yellow or chestnut-brown. Each cyst may contain up to 600 eggs which can remain viable in the soil for many years. There are no chemicals readily available to amateur gardeners for dealing with this pest. Rotation of crops helps to avoid building up a damaging number of cysts in the soil, but if this has already happened it may be necessary to stop growing potatoes for seven or more years. Tomato is also a host plant so it should not follow potatoes in a rotation. Some potato varieties are resistant to *G. rostochiensis* but at the moment there is none that has immunity to *G. pallida*. Resistant varieties include 'Maris Piper' (main crop) and 'Pentland Javelin', 'Pentland Lustre' and 'Pentland Meteor' (earlies). Their roots are attacked by the larvae of *G. rostochiensis* in the usual way but they cannot develop into females and so there is no reproduction. Which species of eelworm is present on a plant can be discovered by examining the roots—if the cysts are all white or brown then the species is probably *G. pallida*, but if there are also some golden yellow cysts then *G. rostochiensis* is more likely to be the species. *G. pallida* can reproduce normally on the above varieties.

Leaves discolored

Magnesium deficiency is a very common problem in wet seasons for potatoes growing in light soils. The tissues between the veins on the older leaves first turn yellow, then brown and become brittle; growth may be stunted. Spray affected plants at the first sign of trouble with $\frac{1}{2}$ lb of magnesium sulfate in 3 gal of water plus a spreader and repeat once or twice at two week intervals.

Stems rotting

Blackleg is due to the bacterium *Erwinia carotovora*. Early in the season it causes the foliage of an affected plant to turn yellow and the shoots to collapse due to blackening and rotting of the stem bases, although occasionally one or two healthy stems develop. The plant may die before any tubers form but any which have already developed show a brown or gray slimy rot inside starting at the heel end. Destroy affected plants. If severely infected tubers are stored they will

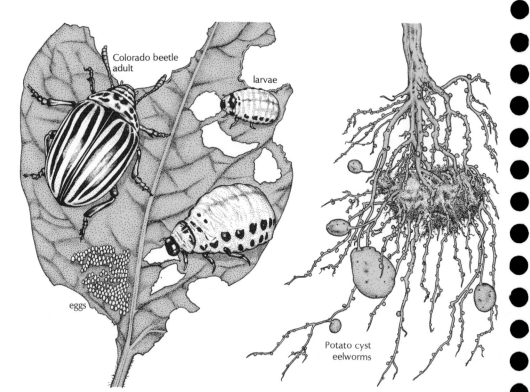

Colorado beetle adult

larvae

eggs

Potato cyst eelworms

decay, but those only slightly infected may show no symptoms and, if planted, will introduce the infection the following season. Only one or two plants in any one crop are likely to show these symptoms since the disease usually occurs as a result of planting such infected tubers. Healthy tubers can be infected at lifting time through direct contact with a diseased tuber. Therefore, once the disease has appeared lift the rest of the crop carefully at harvesting and store only healthy tubers. The disease does not spread from plant to plant in the garden and there is practically no risk of the soil becoming contaminated unless there are large numbers of the bacteria present and the soil is very wet.

Tubers with holes

Small black slugs (*Milax* spp) are the most common cause of holes in potatoes. Comparatively small holes are made in the outside of the potato but the inside is extensively tunneled and it may become completely

hollow. Attacks generally start in August, so early varieties often escape damage as they are lifted before then. In gardens where slugs are known to be a problem, main crop potatoes should be lifted as soon as the tubers have matured in order to limit the extent of the damage. Store holed potatoes separately for early use as they are liable to develop storage rots. No potato varieties are resistant but they do vary in their susceptibility. 'Maris Piper' and 'Pentland Crown' are liable to be badly damaged, while 'Stormont Enterprise' and 'Pentland Ivory' are less severely attacked. The use of organic fertilizers can encourage slugs and should be avoided where a slug problem exists. The types of slug that attack potato tubers spend most of their lives below soil level where they cannot be reached with slug pellets. They do come to the surface, however, during warm weather after periods of heavy rain. Use of methiocarb pellets at such times will help to reduce the slug population.

Potatoes 2

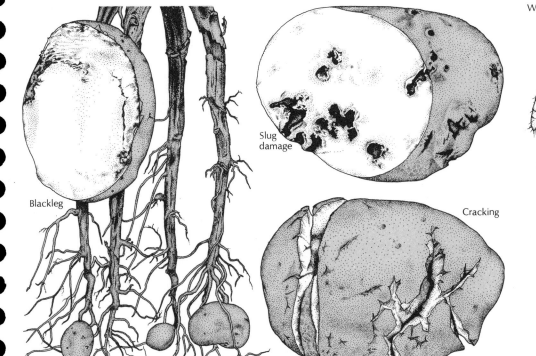

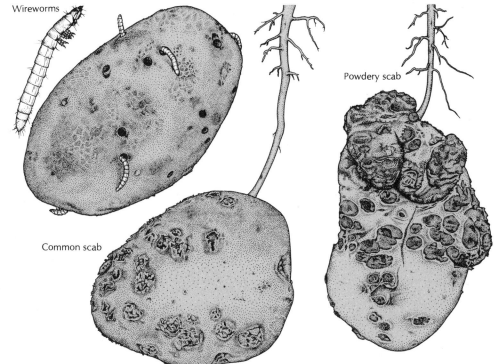

Wireworms (*Agriotes* spp) are the larvae of click beetles. They occur in the greatest numbers under grassland or weed-infested areas and are mainly a problem in new gardens where they can cause damage for three or four years after the land has been brought into cultivation. Wireworms are thin, orange-brown larvae up to 1 in long. They have three pairs of small legs but these stiff-bodied insects do not move very much. The holes they make in the tubers are about $\frac{1}{8}$ in in diameter but, although they may penetrate deeply, the tubers do not become hollowed out as occurs with slug damage. In new gardens it is worth while taking precautions against wireworms and other soil pests such as cutworms and chafer grubs by treating the soil at planting time with any of the pesticides dïazinon, chlorpyrifos or bromophos. If wireworms are present in large numbers, damage to main crop potatoes can be limited by lifting the crop as soon as the tubers have matured.

Large cracks can develop in potatoes which have suffered from drought and have then been watered. Similarly a long wet spell after dry weather can cause large holes in the center of the tubers, a disorder known as hollow heart. These troubles can be prevented by watering early crops at 10–16 day intervals, and main crop potatoes at 4–5 gal per square yard when the tubers are $\frac{1}{2}$ in across. Water regularly in dry weather since it is important that the soil is never allowed to dry out completely.

Tubers scabby
Common scab (*Streptomyces scabies*) causes superficial raised scabs with ragged edges on potato tubers. In severe cases the whole surface may become scabby resulting in considerable wastage when the tubers are peeled for cooking. The organisms causing this disease are found most commonly in soils that lack organic matter, especially dry, sandy or gravelly soils. The texture of such

soils should, therefore, be improved by digging in plenty of humus such as manure, peat, leaf-mold, compost or even mustard seed used as green manure. Do not lime the soil prior to planting as this will encourage the disease. If possible irrigate the soil during a dry growing season to prevent the soil from drying out. Plant only healthy tubers and in severely infected soil grow resistant varieties such as 'Arran Pilot', 'Maris Peer', 'Pentland Crown' and 'King Edward'.

Powdery scab, also known as corky scab, is caused by the fungus *Spongospora subterranea*. In the early stages it shows on affected tubers as scabs which are almost identical to those caused by common scab, but they are generally more circular in outline and have a raised margin. Eventually the scabs burst open to release a brown mass of powdery spores which contaminate the soil. Occasionally a canker form of the disease occurs when affected tubers are so malformed that they may be unrecognizable as

potatoes. The disease is only likely to be really troublesome in a wet season or on low-lying wet soil where potatoes are grown too often, but the variety 'Pentland Crown' is very susceptible to infection and should not be grown in a garden where this disease has occurred. Burn diseased tubers and keep the infected site free of potatoes for three years.

Flesh discolored
Spraing is caused by tobacco rattle virus and, less frequently, by potato mop-top virus. It shows in the flesh of affected tubers as red-brown lesions which have a wavy or arc-like appearance in cross-section. If such tubers are used for seed a mottling of the stems or leaves will appear on the developing plants. Do not plant 'Pentland Dell' where this disease has occurred since it is the most severely affected variety. Tobacco rattle virus is transmitted by certain free-living eelworms present in the soil (which are distinct from the potato cyst eelworm). Therefore always grow

Potatoes 3

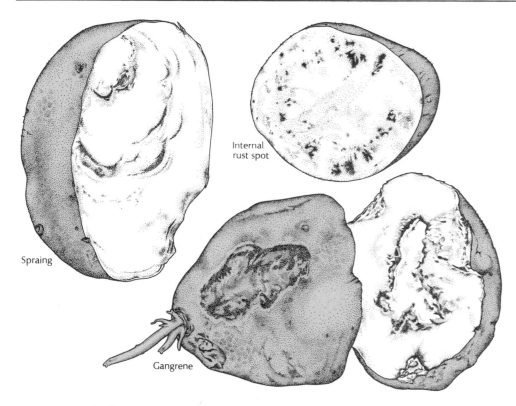

Spraing

Internal rust spot

Gangrene

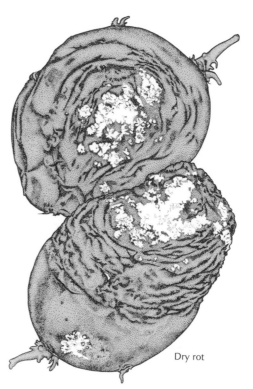

Dry rot

This is caused by the fungus *Phytophthora infestans* and is the most serious disease of potatoes as it can affect leaves, stems and tubers. The first symptom is the development of yellow-brown patches on the leaves. These occur as early as May but the most common time is mid- to late summer. In dry weather the patches become dry and brown but when wet the blotches spread and the leaves and stems turn black and quickly rot. Masses of white fungal threads develop on the under-surfaces of the leaves and produce spores which are wind-borne or washed down by rain into the soil where they infect fresh haulms and tubers near or at soil level. A blighted tuber shows a red-brown dry rot under the skin which spreads inwards. Later, secondary bacteria are likely to enter infected tubers and reduce them to an evil-smelling soft wet mass. Prevent infection of the tubers by planting only healthy tubers in drills at least 5 in deep, and earth up properly. Spray main crop potatoes in early July with manco-zeb, maneb, chlorothalonil, captafol or a copper fungicide. Cut off any green haulms before lifting. Potato varieties resistant to some strains of the blight could be grown in areas where blight is troublesome, for example, 'Maris Peer', 'Pentland Beauty' and 'Record'.

potatoes on a fresh site after an incidence of this disease has occurred.

Internal rust spot is rather like spraing except that the brown marks are scattered within the flesh when an affected tuber is cut open. The exact cause of this trouble is not known but it is thought to be a physiological disorder: symptoms are less severe in a crop where even growth is maintained throughout the season by preventing any checks due to sudden drying out of the soil. Dig plenty of humus into the soil before planting and in dry seasons irrigate the crop when possible. Grow resistant varieties such as 'Arran Consul' and 'King Edward' if internal rust spot is a recurrent problem.

Blackening of the flesh commonly occurs after cooking in many varieties including 'Home Guard', 'Majestic', 'Pentland Dell' and 'Ulster Chieftain'. Sometimes however, the blackening occurs before cooking. These troubles are usually due to a deficiency of potash during the growing season or storing

potatoes at too high a temperature, especially if they are kept in unventilated bags.

Tubers rotting

Gangrene is a storage disease which first shows as a slight discolored depression bearing minute black fruiting bodies of the fungus. If the storage conditions are poor the diseased area enlarges until most of the tuber is decayed and shrunken. If it is cut open at this stage it will be hollow and rotting inside. Gangrene is caused by the fungus *Phoma solanicola* f *foveata* which is probably present in most soils and may occur on healthy tubers. It normally only affects tubers which have been injured, particularly when lifting and especially if the soil is very wet at the time. Be careful therefore when lifting potatoes so that they are not injured, and store them in proper storage conditions. Burn severely rotting tubers but with expensive seed tubers cut out the diseased areas and plant the healthy parts.

Dry rot (*Fusarium* spp) is a disease of stored seed tubers which develops from January onwards. The first symptom is a shriveling of the skin at one end of the tuber, followed by a rapid shrinking of the affected tissues. White or pink pustules of fungus spores develop on the shrunken tissues and, as the disease extends, the rotting tuber is reduced to a shapeless mass that is useless for planting. The organisms that cause dry rot are present in all soils and infection occurs while the tubers are still in the ground or at lifting time, the fungus entering through eyes, breathing pores, scab wounds or abrasions. Prevent wounds by treating seed potatoes carefully at lifting or when storing, which should be done in a frost-proof but cool, dry and well ventilated place as the disease is favoured by damp, warm (16°C/61°F) airless conditions. When buying seed potatoes take them out of the bags on delivery, remove diseased tubers and place the others in the sprouting-trays.

Carrots and parsley

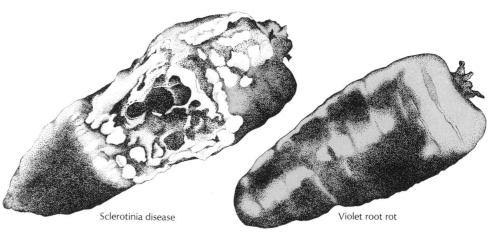

Motley dwarf disease

Carrot fly adult

maggot

Splitting

Sclerotinia disease

Violet root rot

Leaves discolored

Carrot motley dwarf disease is caused by a combination of the carrot motley dwarf and carrot red leaf viruses. Affected plants are stunted with twisted leaf stalks. The foliage shows yellow bands between the veins at first, and then becomes highly colored with red tints. Individual leaflets become distorted and fail to develop, and the root tips and root hairs die. Similar symptoms appear on parsley affected by either or both of these viruses. Dig up and burn affected plants, since there is no cure. Prevent infection by spraying with dimethoate to control aphids, since these insects may transmit the viruses.

Roots tunneled

Carrot fly (*Psila rosae*) is the most damaging pest of carrots, and it also attacks related plants such as parsley, parsnips and celery. There are two generations of the maggots during the summer with damage occurring from June to July and from August to October. The maggots are thin, orange-white larvae up to $\frac{1}{2}$ in long. They tunnel just below the surface of the carrot root causing a rusty-brown discoloration and giving the carrot an unpleasant taste. Attacks on young plants can stunt their growth; attacks later in the season can allow secondary rots to develop which cause the carrot to decay in the soil or during storage. Sowings made after the end of May will miss the first generation of maggots, while early sowings will be ready

for eating before the second generation appears in August. However, this pest is so widespread and damaging that at least one application of insecticide will be necessary. When sowing, treat the seed drill with diazinon or chlorpyrifos granules. This will protect the plants for about six to eight weeks. Carrots that are not going to be lifted until the fall should also be watered thoroughly with spray-strength trichlorphon in late August. Egg-laying female flies locate suitable host plants by scent and are strongly attracted to recently thinned rows because the process of thinning bruises the foliage and releases the attractive odor. Reduce the need for thinning by careful spacing and using pelleted seed.

Roots splitting

Faulty root action is usually due to an irregular supply of moisture in the soil and can result in the roots splitting lengthwise. Maintain even growth throughout the season by mulching to conserve moisture and watering the soil before it dries out completely. Do not store cracked roots as they will not keep.

Roots rotting

Sclerotinia disease, caused by the fungus *Sclerotinia sclerotiorum*, is the most destructive disease of carrots. Plants can be attacked during the growing period but most commonly it causes storage rot. White fluffy fungus growth develops on the rotting

roots and hard black structures $\frac{1}{2}$ in or more in length soon form on the diseased tissues. These are the resting bodies of the fungus. During the growing season destroy any plants which show rotting shoots covered by a white fluffy fungus, preferably before the resting bodies form, otherwise they fall off into the soil and contaminate it. They remain dormant during the winter and germinate the following spring to give rise to spores which attack first dead, then living, tissues. Store only healthy roots in a well ventilated and fairly dry place. Check them regularly and remove and burn any rotting roots before the resting bodies are produced and fall to the floor, thus contaminating the storage area. Dusting seed with dry benomyl before sowing may help to prevent infection.

Violet root rot (*Helicobasidium purpureum*) attacks the roots while still in the soil. The fungus grows over the surface of the carrot forming a violet or purple web of fungal threads, and the tissues shrink and rot. Destroy affected carrots and do not grow them, any other susceptible root crops or asparagus on the site for several years.

Soft rot (*Erwinia carotovora*) can occur on carrots that have been stored incorrectly. Affected carrots become soft and slimy, and develop a foul smell. The bacteria can only enter through wounds so do not store roots damaged by carrot fly or any other injury. See that storage conditions in the clamp or shed are not too moist.

Parsnips/Annual spinach

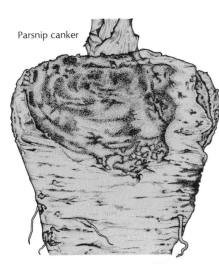

Parsnip canker

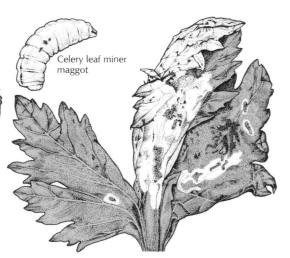

Celery leaf miner maggot

Carrot fly maggot

Spinach blight

PARSNIPS
The most important pests and diseases of parsnips are those that affect the roots.

Roots splitting
Drought followed by heavy rain or irrigation can cause the roots of parsnips to split. Prevent by growing in ground that was manured for a previous crop, since the humus will help to conserve moisture, and by watering before the soil dries out completely.

Roots rotting
Parsnip canker is caused mostly by the fungus *Itersonilia perplexans*. Unsuitable soil conditions may also result in cankering of the tissue associated with horizontal growth cracks. In general, canker is any form of rotting of the shoulder tissues of parsnip during the fall and winter. Shallow and superficial red-brown or black lesions develop but in severe cases the rot extends fairly deeply into the tissues both at the crown and on the shank. There are no satisfactory control measures. The incidence of the disease can be reduced by growing parsnips on a fresh site each year in deep, well worked soil to which is added a balanced fertilizer. Add lime if the pH of the soil lies below 6.5–7.0. Sow the seed early and thin to leave the plants at intervals of 3 in instead of the usual 6 in. Grow resistant varieties such as 'Avon-resister' and 'Tender and True'.

Roots with pests visible
Root aphid (*Anuraphis subterranea*) is a gray-white aphid that forms dense colonies on parsnip, especially on the shoulders of the root and at the base of the leaves. The aphids are most frequently seen in late summer. Control them by thoroughly spraying the base of the plants with a systemic insecticide.

Roots tunneled
Carrot fly maggots (*Psila rosae*) are less damaging to parsnips than carrots but they can nevertheless spoil the appearance of the roots and allow rots to enter through the wounds. The slender creamy-white maggots, which grow up to ½ in long, feed by tunneling below the surface of the root. These shallow tunnels turn rust-brown and are readily seen when the roots are being peeled. Protect young plants by treating the seed drill with diazinon or chlorpyrifos granules at the time of sowing. There are two generations of carrot fly maggots during the summer. Protect the plants against the second generation by watering with spray-strength trichlorphon in late August.

Leaves mined
Celery leaf miner maggots (*Euleia heraclei*) live inside the leaves, causing pale green blotched areas to develop which later become brown and dry. Several white maggots occur inside each mine. There are two gener-

ations with damage occurring in April to June and in July to September. The first generation is the more damaging as it can stunt the growth of the root. Control light infestations by picking off the affected parts of leaves. Spray extensive attacks with trichlorphon or dimethoate.

SPINACH
The following diseases may afflict annual spinach. For pests and diseases of spinach beet, see page 48.

Leaves spotted
Leaf spot is caused by the fungus *Heterosporium variabile*. It shows as circular light-brown or gray areas with well defined brown or purple edges. The spots may increase rapidly on weak plants. Remove and burn affected leaves when the first symptoms are seen. Encourage the plants to grow strongly by spraying with a foliar feed. To prevent subsequent attacks apply a dressing of potash at ¼ oz per square yard before sowing on a fresh site, and thin out early.

Leaves discolored and distorted
Spinach blight due to cucumber mosaic virus shows first as a yellowing of the younger leaves and later of the older ones. As the older leaves die just a few distorted central leaves remain, being much narrower and puckered with inrolled margins. Control

aphids, which spread this disease, by the use of systemic insecticides. Keep weeds down since these can harbor the virus, and remove and burn all diseased plants.
Manganese deficiency produces yellow blotches between the veins of the leaves, which have a tendency to roll inwards at their edges. In severe cases whole leaves become pale yellow. Where this is a recurrent problem, spray with a solution of manganese sulfate at 2 oz to 3 gal of water plus a spreader, repeating once or twice at two week intervals. Alternatively, rake in either fritted trace elements or a chelated compound before sowing annual spinach.

Leaves moldy
Downy mildew (*Peronospora effusa*) produces a gray or violet mold on the lower surfaces of leaves and yellow blotches on the upper surfaces. Later, affected areas die and dry out to form light gray-brown patches. The inner leaves usually show only small spots but the outer leaves may stop growing and curl downwards at the edges or be destroyed completely, particularly on crowded plants growing in moist conditions. Avoid overcrowding and encourage circulation of air by proper thinning of plants, which should be grown on a well drained site. Spraying is difficult because of the closeness of the leaves but applications of mancozeb, zineb or a copper fungicide may check the disease.

Rutabagas and turnips

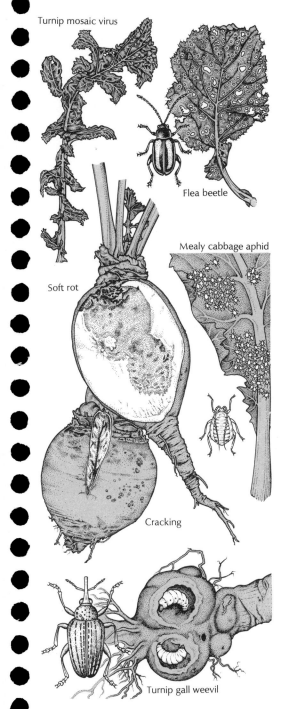

Turnip mosaic virus

Flea beetle

Mealy cabbage aphid

Soft rot

Cracking

Turnip gall weevil

Leaves discolored

Turnip mosaic and cauliflower mosaic are the most common viruses to affect turnips and rutabagas. The symptoms they cause are variable but in general the leaf-veins lose their green color so that they stand out as pale lines on leaves which are twisted and stunted. Mottling may occur, with irregular raised patches of dark green tissue developing among paler areas to give a blistered appearance. Small yellow or brown patches may develop between the veins. Plants infected when young often die. The leaves of older plants die down prematurely, thus exposing the crowns which are then often attacked by soft rot. Control aphids, which can spread the viruses, by using a systemic insecticide. Destroy all affected plants.

Leaves with holes

Flea beetles (*Phyllotreta* spp) attack the seedlings producing many small holes in the leaves. Heavy attacks may kill the seedlings, especially if growth is retarded by poor growing conditions. The beetles are up to $\frac{1}{10}$ in long and are either shiny black or black with yellow stripes. Seed rows should be watered during dry spells to enable the seedlings to grow rapidly through this susceptible stage. If damage does occur, the beetles can be controlled by dusting the row with derris or pirimiphos-methyl.

Leaves with pests visible

Mealy cabbage aphid (*Brevicoryne brassicae*) forms large, closely packed colonies on the underside of swede leaves. The aphids are covered in a gray-white waxy powder. Their feeding causes the leaf to turn yellow. This aphid can be present from May to October, but is most harmful earlier in the year as infestations can then permanently stunt the growth of young plants. Rutabagas and other brassicas should be inspected regularly during the summer for aphids. Control them by spraying with a systemic insecticide.

Roots rotting

Soft rot is caused by the bacterium *Erwinia carotovora*. It can occur on stored roots and can also attack plants during the growing season, particularly in wet weather when many plants may be destroyed. The bacteria can enter through wounds in the roots, but more generally they affect the crown or bases of the leaf stalks causing the stem to become soft and slimy and the leaves to drop. The flesh of infected roots is soon broken down into a white or gray soft, rotting and unpleasant-smelling mass, but the rind remains intact and affected roots often become hollow. The disease is most troublesome on crops grown in heavily manured and badly drained soil in gardens where little rotation is carried out. Rotation of crops, adequate drainage, careful manuring and the avoidance of injuries to the plants during cultivation should prevent the disease. Control virus diseases and pests such as slugs which may injure the foliage or cause it to die down early thus exposing the crowns. Remove affected plants and those nearby immediately and burn all infected material. Store only healthy roots in a dry frost-proof shed or in a well made clamp.

Roots galled

Club root (*Plasmodiophora brassicae*) distorts the roots below the edible part of rutabagas and turnips which become swollen and gnarled. If cut open the flesh of the gall is somewhat mottled. Affected plants can become stunted and may wilt. Since the disease occurs in acid soil, raise the pH level by applying ground chalk or limestone at 14 oz per square yard, or hydrated lime at 10 oz per square yard. In succeeding years 8 oz per square yard should be sufficient. Improve the drainage by digging the soil deeply and by incorporating plenty of humus. Rotate crops so that brassicas and other crucifers, including weeds, are kept off infected land for as long as possible. If club root disease becomes a recurrent problem apply the fungicide PCNB to the furrows when young rutabaga or turnip plants are being transplanted to their final positions.

Turnip gall weevil (*Ceutorhynchus pleurostigma*) causes marble-like swellings on the roots of turnips, rutabagas and brassicas. These can be mistaken for club root disease but may be recognized by cutting the swellings in half. Weevil galls are hollow, and either contain the weevil's grub or show a circular hole in the side of the gall where the grub has left to pupate in the soil. Club root swellings are solid, less regular in shape and frequently smell of rotting cabbages. Weevil galls spoil the appearance of the roots but they do not seem to affect the plant's growth. In gardens where this pest occurs, plants can be protected by applying diazinon granules to the furrows at the time of sowing. There is only one generation of weevils a year, and this treatment is usually effective in preventing attacks. If damage still occurs, apply diazinon, chlorpyrifos granules or bromophos powder to the seed drill.

Roots with discolored flesh

Brown heart is due to a deficiency of boron. Affected plants show no external symptoms and the trouble is only noticed when their roots are cooked as they are then hard, stringy and tasteless. However if an affected root is cut across the flesh in the lower part it will show clearly defined gray or brown discolored areas, often in concentric rings. Prevent this trouble by applying borax (sodium tetraborate) to the soil before sowing. It is applied at the rate of 1 oz per 20 sq yd, but should be mixed with a large amount of light sand to make even spreading easier.

Roots split

Heavy rain or watering following a long, dry spell may cause the roots to split. Prevent by digging in plenty of humus to conserve moisture, by mulching and by watering in dry periods before the soil dries out completely. Do not store cracked roots.

Roots eaten

Cabbage maggots (*Hylemya brassicae*) may destroy the feeding roots of rutabagas and turnips, and also tunnel into the swollen part of the roots. Damaged plants tend to grow slowly, have discolored foliage, and are liable to wilt in sunny weather. When such plants are dug up, the lack of feeding roots can be seen and white, legless maggots up to $\frac{3}{8}$ in long may be found amongst the remains of the roots. Once plants get to this stage it is really too late to save them. Damage can be prevented by treating the seed rows with diazinon, chlorpyrifos or bromophos.

Beetroot and spinach beet

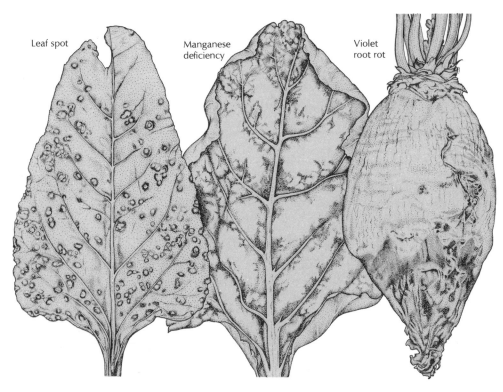

Leaf spot

Manganese deficiency

Violet root rot

Beet leaf miner maggots

Leaves discolored

Magnesium deficiency shows on the older leaves of purple-leaved beet as pale areas between the veins followed by browning of the tissues. Green-topped varieties of beet and spinach beet may develop bright red tints. At the first sign of trouble spray with a solution of magnesium sulfate at a rate of $\frac{1}{2}$ lb in 3 gal of water to which is added a spreader. Give two or three applications at two week intervals.

Manganese deficiency causes an inward rolling of the edges of the leaves which become triangular in outline and show yellow blotches between the veins. This condition is known as speckled yellows. In extreme cases whole leaves become pale yellow. Where this occurs annually spray with a solution of manganese sulfate, 2 oz in 3 gal of water plus a spreader, two or three times at two week intervals. Alternatively, rake in fritted trace elements or a chelated compound before sowing.

Leaves spotted

Leaf spot can be caused by the fungi *Cercospora beticola* and *Ramularia beticola* and shows as brown spots on the leaves. The diseased tissues often fall away leaving holes. Severe symptoms are only likely to occur on soft-grown plants or those growing too close together, particularly in wet weather. Prevent leaf spot by growing on a fresh site each year and by applying a dressing of potash at $\frac{1}{4}$ oz per square yard before sowing. Thin out early and well. Should the disease occur, remove affected leaves immediately and spray plants with a solution of zineb at ten day intervals.

Leaves mined

Beet leaf miner maggots (*Pegomya hyoscyami*) live together in small groups inside the foliage and cause irregularly blotched mines. These are at first yellow-green, but later turn brown as the damaged areas dry up and shrivel. Damage occurs between May and September. Young plants can have their growth checked, so early attacks are more harmful than those that occur when the root is fully grown. Light infestations can be controlled by picking off the affected leaves or by crushing the maggots inside the mines. If insecticides are necessary the plants can be sprayed with trichlorphon or dimethoate.

Roots splitting

Heavy rain or watering after a long, dry spell can cause beetroots to split. Prevent this trouble by incorporating plenty of humus to conserve moisture and by watering in dry periods before the soil dries out completely. Do not add fresh manure as this may cause forking of the roots.

Roots with unhealthy skin

Scab (*Streptomyces scabies*) occurs occasionally, and then usually shows as only slight external blemishes. Sometimes however numerous sunken scabby pits or knob-like raised scabs appear. The scabs may be dotted about singly or develop in patches at approximately the same level on the root, a condition known as girth scab. Although unsightly this disease does not affect the quality of the roots. However, since this disease is worse in light soils lacking organic matter, dig in plenty of humus such as peat, leaf-mold or compost before sowing, and do not lime. If possible irrigate in dry periods to ensure that the soil does not dry out.

Violet root rot (*Helicobasidium purpureum*) can attack beetroots if they are grown in infected soil. The leaves may become slightly yellow or stunted during the growing period, but the disease is not seen until lifting when the web of violet or purple strands of fungus will be seen on the roots. Burn affected roots. Do not grow asparagus and susceptible root crops such as potato, beet, turnip, swede or carrot on the infected site for several years, and keep it free of weeds since these may also harbor this rot.

Roots discolored internally

Heart rot is due to a deficiency of boron. A dry rot, it shows as a browning of the inner tissues of the root, which sometimes turns black, and also at the crown where the tissues may become sunken. In severe cases the outer tissues of the root may also rot causing cankers. Most of the leaves die and only small deformed leaves remain. Where this is a recurrent problem rake borax (sodium tetraborate) into the soil before sowing, applying it at 1 oz for every 20 sq yd, first mixing it with sufficient light sand to make spreading easier.

Flowers formed

Bolting shows as the development of flower stalks during the year of sowing. It is caused by a check in growth such as will occur if thinning is carried out too late or the soil dries out completely. Sowing too early may also cause plants to run to seed before the roots have become large enough to eat. Prevent drying out of the soil by digging in humus, but not fresh manure, before sowing and water in dry periods. Sow at the correct time and thin early. The variety 'Boltardy' is resistant to bolting.

Rhubarb/Sweetcorn

Rhubarb

Leaf spot

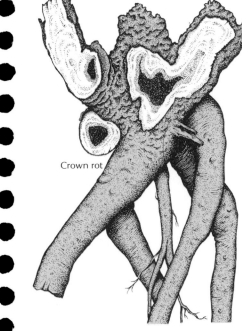

Crown rot

Leaves discolored

Faulty root action due to very wet or dry soil conditions results in brown or yellow blotching of the leaves or general discoloration of the foliage. If waterlogging occurs frequently drain the site or replant in a raised bed. In dry periods water before the soil dries out completely. If the symptoms appear very early in the season, spray with a foliar feed to help the plants overcome the check in growth.

Leaves spotted

Leaf spot (*Ramularia rhei*) shows as irregular-shaped brown spots all over the leaves. Diseased areas may fall away leaving holes. It is of little consequence and does little damage to the plant. Nevertheless, diseased leaves should be removed and burned. If the disease is severe the plant is probably lacking in vigor due to faulty root action and would benefit from applications of a foliar feed.

Crowns rotting

Crown rot (*Erwinia rhapontici*) is a soil-borne disease that is particularly troublesome in wet soils. The bacteria commonly enter through wounds and cause rotting of the terminal bud, together with a soft brown rot of the inner tissues below the crown, forming a cavity within the pith. Affected crowns deteriorate gradually until only a few small splindly sticks are produced by the lateral buds; these have a dull color and may also rot. Burn diseased plants and plant new healthy roots on a fresh site, which should be well drained.

Honey fungus (*Armillaria mellea*) is only likely to be troublesome if rhubarb is grown in soil that is already infected with this root parasite, such as old orchard sites. Affected plants die and show white streaks of fungal growth within the dead tissues of the root and crown, which may also suffer from a mushy rot. The fungus spreads through the soil by brown root-like structures called rhizomorphs which may be attached to the rotting roots. Dig up affected plants together with all the roots and burn them. Grow annual vegetables for the next few years, or change the soil completely. Alternatively sterilize the soil with a 2 per cent solution of formalin before replanting with fresh stock.

Sweetcorn

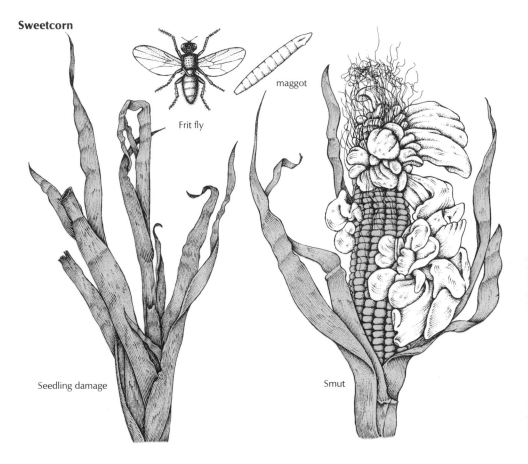

Frit fly

maggot

Seedling damage

Smut

Seedlings distorted

Frit fly (*Oscinella frit*) lay their eggs on seedlings. The small white maggots feed inside the shoots and can cause the seedlings to die. Several side-shoots are subsequently produced, thus resulting in delayed and weakened growth. The young leaves have a tattered and distorted appearance where they have been eaten by the maggot. Once the seedlings have grown beyond the five-leaf stage they are no longer attractive to the egg-laying female flies. Seedlings can be grown through the susceptible stage in pots in a greenhouse where they are unlikely to be found by frit flies. Alternatively, if sweetcorn is to be raised from seed in the garden, irrigate as often as necessary to keep the seedlings growing rapidly until they are no longer liable to be attacked. In particular, ensure that the soil does not dry out.

Leaves, stems and cobs with swellings

Smut (*Ustilago maydis*) is only troublesome in long, hot summers. "Smut balls", which are up to 5 in in diameter and composed of a white, smooth covering of corn tissue, develop on the ears, tassels and the joints of the stem. Each ball encloses a mass of black greasy or powdery spores which are released as a dust when the covering becomes brittle and breaks open at maturity. Examine plants regularly during a hot summer and cut off and burn any balls before they burst open. Burn all debris at the end of the season and do not grow sweetcorn on the site for several years. Alternatively, grow resistant varieties.

Outdoor tomatoes 1

This section is concerned only with tomatoes grown out of doors. For greenhouse tomatoes, see Protected crops, pages 54–6.

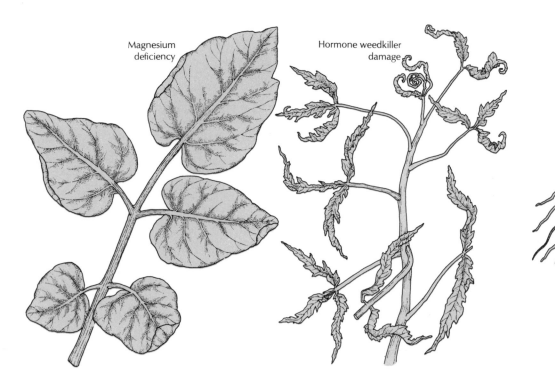

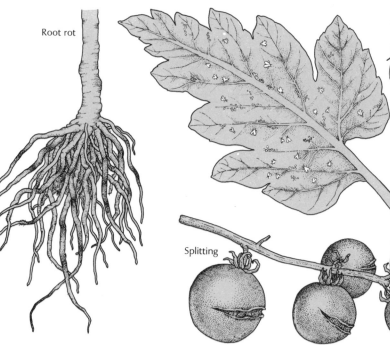

Leaves discolored

Magnesium deficiency is very common on plants fed with high potash fertilizers, the symptoms being orange-yellow bands between the veins. The lower leaves are affected first and as the symptoms spread progressively upwards the affected areas on the older leaves turn brown. When the yellowing commences spray with a solution of magnesium sulfate made at the rate of $\frac{1}{2}$ lb in 3 gal of water to which is added a spreader. Repeat at seven to ten day intervals throughout the season. Even with this treatment some yellowing of the foliage may persist but affected plants can still produce good crops of high quality fruit.

Leaves distorted

Hormone weedkiller damage due to the misuse of such weedkillers as 2,4-D and mecoprop commonly occurs on tomatoes. It causes distortion of the stems and leaf stalks which twist spirally. The leaves are also affected, becoming narrow and twisted or cupped and showing parallel veining. The later-developing foliage on an affected plant is usually normal but the fruits formed at the time of damage are plum-shaped and often hollow and seedless; they are edible but not necessarily palatable. Prevent such damage by taking precautions when using hormone weedkillers; they should be applied only on a still day using equipment that is kept specifically for their application.

Leaves rolled upwards

Cold nights can cause tomato leaves to roll tightly upwards, especially when there are great extremes between day and night temperatures. The leaves are inclined to become brittle and may be damaged if touched but this symptom is nothing to worry about as it indicates that the plants are healthy and growing well. The leaves remain curled for the rest of the season and there is nothing which can be done either to prevent the trouble or to uncurl them. Affected plants produce good crops of edible fruit.

Leaves with pests visible

Aphids (various species) are not as much of a problem on outdoor plants as in a greenhouse, but they are nevertheless an important pest as they can spread virus diseases. Aphids can have various colors such as green, yellow-green or pink, and they feed by sucking sap from the stems and the undersides of leaves. Heavy infestations cause the leaves to become sticky with the aphids' excretions, and the foliage and fruits may be blackened by a sooty mold that grows on this honeydew. Control them by spraying with pirimicarb or a systemic insecticide. If the pest is not noticed until the tomatoes are ready for eating, a short-persistence chemical such as a pyrethroid or derris should be used.

Greenhouse whitefly (*Trialeurodes vaporariorum*) is, like aphids, mainly a greenhouse problem but it can attack outdoor plants, especially in hot summers. The adults are small white moth-like insects that fly up from the undersides of the leaves when the plant is disturbed. Both the adults and their scale-like larvae suck sap from the leaves, which become soiled with a sticky honeydew and a black sooty mold. Control measures should be taken as soon as whitefly is seen since heavy infestations are difficult to control. Suitable sprays are pyrethroid compounds such as pyrethrum, resmethrin and bioresmethrin, or other chemicals such as pirimiphos-methyl. Use one of the pyrethroids if the fruits are ready for picking. Spray plants thoroughly at weekly intervals until the pest has been controlled.

Stems wilting

Foot and root rot can be caused by numerous fungi including *Thielaviopsis basicola*, *Rhizoctonia solani*, *Aphanomyces cladogamus* and species of *Phytophthora* and *Pythium*. Rotting of the roots or at the base of the stems leads to discoloration of the foliage and often complete collapse of the plant. Tomato plants that are over- or under-watered or have been poorly planted are susceptible to infection. Plant out carefully on a fresh site

Outdoor tomatoes 2

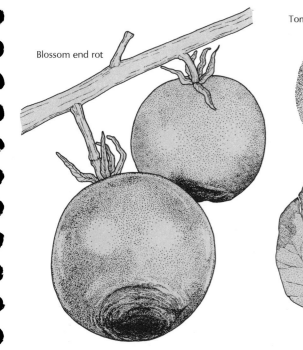

Blossom end rot

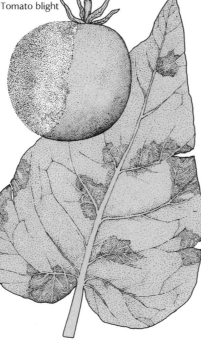

Tomato blight

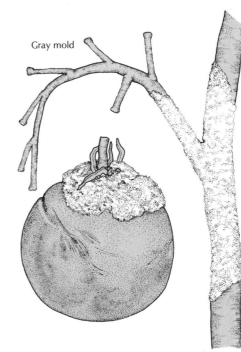

Gray mold

VIRUS DISEASES

Many different viruses can attack tomatoes, the most troublesome being tobacco mosaic virus. These viruses can infect tomatoes separately or in combination. Symptoms include stunting of plants, distortion of foliage, a dark green or yellow mottling or spotting of the leaves, dark streaking of leaf stalks and stems, bronzing of leaves and fruit, and brown pits on the fruit with brown markings in the flesh beneath the skin. Destroy any plant immediately it shows any of the above symptoms and wash hands and tools with soapy water before handling or trimming the remaining healthy plants. Grow tomatoes on a fresh site the following season since tobacco mosaic virus can remain on infected tomato debris present in the soil.

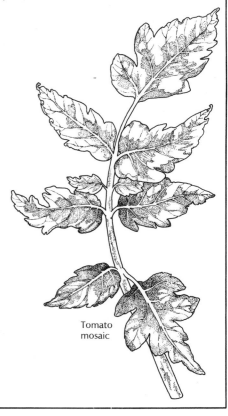

Tomato mosaic

each year in well prepared soil and look after the plants well. Water carefully but make sure the soil is not allowed to dry out.
Soil-borne fungi (*Fusarium oxysporum* f *lycopersicae* and *Verticillium albo-atrum*) cause tomato plants to wilt and die at the time of fruit production. Prevent these diseases by planting only those varieties certified to be "f" and "v" resistant.

Fruit splitting
Irregular watering may cause the development of rings near the shoulder of the fruit or lengthways cracks. Prevent the soil from drying out by digging in plenty of humus to conserve moisture, by mulching and by watering in dry periods. Remove splitting tomatoes early before they become attacked by secondary infections.

Fruit with a dark patch
Blossom end rot shows as a circular brown or green-black patch, often sunken, at the end of the fruit farthest away from the stalk. It is

infrequent on outdoor crops but can occur if watering is irregular. Blossom end rot is caused by a deficiency of calcium within the fruits, but this is rarely due to a lack of calcium in the soil except in very acid soils found in some growing bags. Prevent the trouble by seeing that the soil is never allowed to dry out completely, particularly if the plants are in growing bags.

Fruit rotting
Gray mold fungus (*Botrytis cinerea*) can be troublesome on outdoor tomatoes in a wet season. A brown-gray velvety growth of fungus develops on rotting fruit and also on leaves and stems. The first sign of trouble however may be pinpoint spots with pale-colored rings on the green fruits, a condition known as water spot or ghost spot. Fruits showing this symptom do not usually rot. Spores of the fungus are always present in the air and infect plants through wounds and dead and dying tissues. Good hygiene such as the prompt removal of dead leaves

and over-ripe fruits will help to prevent this trouble. Apart from removing and burning diseased parts at the first signs of disease no other measures are usually necessary on outdoor tomatoes.

Tomato blight (*Phytophthora infestans*) can be very troublesome in wet weather in August and September. It shows first as brown spots or patches on the leaves with sparse white fungal growth on the lower leaf surfaces. An affected leaf tends to dry up and curl but the disease soon spreads to the stems which show blackened patches and may collapse as a result of the rot. Attacked fruits also become brown then shrink and rot rapidly. Fruit apparently healthy when picked from unsprayed plants may develop the rot about five days later. Spray plants against blight every season, applying a copper fungicide, mancozeb, or captafol as soon as the tops have been pinched out of most of the plants. In cool wet seasons repeat applications of these fungicides at two week intervals.

Cucurbits 1

This section covers outdoor cucumbers, marrows, zucchini, pumpkins and squashes. For cucurbits grown under cover (including melons), see Protected crops, pages 54–6.

Greenhouse whitefly

Cucumber mosaic virus

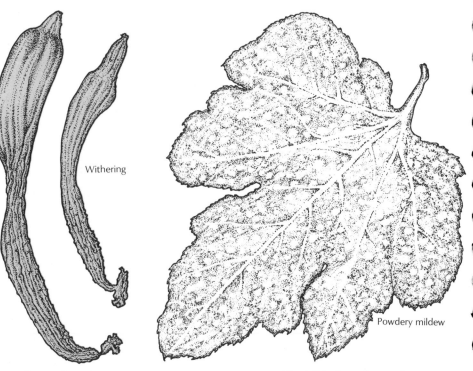

Withering

Powdery mildew

Leaves distorted

Cucumber mosaic virus commonly affects cucurbits, and is particularly troublesome on marrows. Diseased plants are stunted and bear puckered distorted leaves which are mottled with yellow or light and dark green patches. Destroy any plant immediately it shows these symptoms. Failure to do so allows the disease to spread to the fruits (see below), and other plants may also become infected. The virus has a wide host range including weeds, which should therefore be controlled. Spray against aphids since they spread the virus. Whenever possible, plant only those varieties of cucurbits that are listed as being CMV or WMV resistant. The virus is frequently spread by tools and on hands when diseased plants are handled before healthy ones. When handling plants leave the diseased ones until the end before pulling them out and burning them. Alternatively, wash hands and tools with soapy water after handling diseased plants to prevent the virus from spreading.

Leaves with visible fungal growth

Powdery mildew (*Erysiphe cichoracearum*) commonly occurs on marrows and occasionally on outdoor cucumbers, particularly on plants which are dry at the root. The leaves and stems become covered with a white powdery coating of fungus spores. Spray with benomyl, dinocap, sulfur or chlorothalonil when the first symptoms appear and repeat as necessary. Reduce the possibility of infection by careful watering to prevent the soil from drying out completely.

Leaves with pests visible

Aphids (various species), also known as greenfly, damage the plants by sucking sap from the leaves and shoot tips. They not only weaken the plant but also spread virus diseases which can make the plant completely unproductive. Members of the cucumber family are sensitive to some insecticides, so care has to be taken in their use. If the soil is dry, always water the plants before spraying, which should be done at

Cucurbits 2

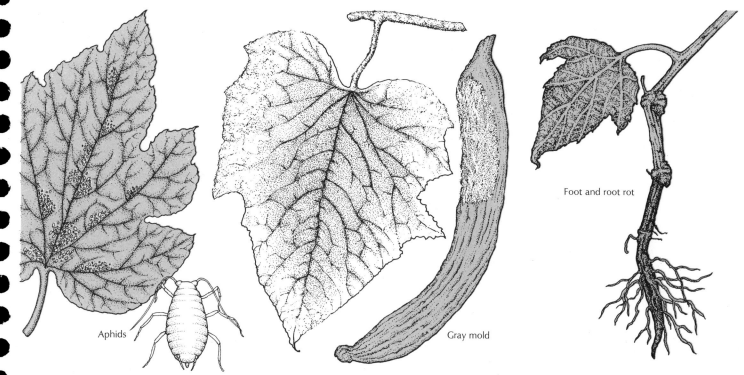

Aphids

Gray mold

Foot and root rot

dusk when temperatures are lower. Pyrethroid compounds such as resmethrin, bioresmethrin and pyrethrum are safe to use on cucurbits if these precautions are taken.

Greenhouse whitefly (*Trialeurodes vaporariorum*) can spread from the greenhouse to garden plants during the summer. Adult whiteflies are small, white, moth-like insects which fly up from the plant when it is disturbed. The flat scale-like larvae are pale green and, like the adults, they can be found on the underside of the leaves where they feed by sucking sap. Follow the control measures described above for aphids. Several sprayings may be necessary at seven day intervals since the eggs and immature stages of this pest, but not the adults, are relatively immune to insecticides.

Stems wilting

Foot and root rot can be caused by the fungi *Thielaviopsis basicola*, *Rhizoctonia solani* and species of *Fusarium*, *Phytophthora* and *Pythium*. The symptoms are rotting of the tissues at the base of the stem or the roots, causing discoloration of the foliage followed by the complete collapse of the plant. It commonly occurs where the cucurbits are grown on the same site each year. Therefore grow the plants on a fresh area each season and if possible grow varieties that are certified to be fusarium resistant. Water carefully but make sure the soil is not allowed to dry out, since the disease is common on plants which are over- or under-watered. Check the disease on slightly affected plants by watering with a solution of captan, Cheshunt compound or zineb, or dusting the base of the plant with dry bordeaux powder in the case of foot rot. Burn severely affected plants since they will not recover.

Stems rotting

Sclerotinia disease (*Sclerotinia sclerotiorum*) may attack cucumber plants near ground level if grown in soil infected with this fungus. The rotting stems become covered with a white fluffy growth of fungus in which are embedded hard black resting bodies of the fungus $\frac{1}{2}$ in or more in length. These structures may also be found within the rotting stems if they are broken open. Any plant showing these symptoms should be destroyed, if possible before the resting bodies are formed as they fall off into the soil very quickly and remain there until the following spring. They then germinate to produce spores which attack newly developing susceptible host plants. Do not grow cucumbers on the affected site for several years.

Gray mold (*Botrytis cinerea*) can cause severe losses in a wet season, although it is most troublesome under glass. A gray-brown velvety growth of fungus develops on rotting stems, fruits and leaves. Infection can also occur by contact between diseased and healthy fruit, leaves and stalks so it is essential to remove and burn all rotting parts immediately. The removal of dead leaves and over-ripe fruits will help to prevent the trouble. Avoid injury to the fruits caused by over-watering with cold water.

Fruits distorted

Cucumber mosaic virus can cause severe symptoms on the fruit as well as on the leaves. Fruits become distorted and mottled, or turn light green and develop spots or dark green, raised warts. However an infected plant would have shown symptoms on the leaves before this stage was reached and should have been destroyed (see above).

Fruits withering

Faulty root action can result in withering of the young fruits starting at the blossom end. Maintain even growth by careful watering and check that the plant has not been affected by a foot or root rot. If no obvious disease is present, rest the plant by removing the fruits and spray with a foliar feed if the leaves are a poor color. Once the plant has regained vigor the later-developing fruits should be normal.

Fruits bitter

Excess nitrogen in the soil can cause bitterness of cucumbers. Prevent this by avoiding the excessive use of nitrogen fertilizers. Bitterness can be also due to irregular growth so try to maintain even growth by watering carefully to prevent waterlogging or complete drying out of the soil. Outdoor cucumbers are also inclined to become bitter when old.

Fruits rotting

Gray mold (*Botrytis cinerea*) is the only organism which is likely to rot outdoor cucumber and marrow fruits. For symptoms and treatment, see above.

Fruits lacking

Bad weather may cause poor pollination due to a lack of insect activity. This results in the fruit failing to set and is a common problem with marrows and zucchini. Marrow flowers must be pollinated to produce fruit, so if insects are lacking do this by hand. The male flowers are identified by a thin stalk behind the petals; the female flowers have a tiny marrow behind the petals. Pick the male flowers on a dry day in the morning and dust the pollen from them into the center of each female flower. Do not allow the soil to dry out completely.

Protected crops 1

Seedlings

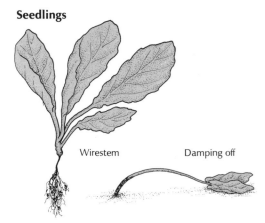

Wirestem   Damping off

Leaves discolored

Downy mildew of brassica seedlings, especially cauliflowers, is caused by the fungus *Peronospora parasitica* and that of lettuce by *Bremia lactucae*. White mealy or downy tufts of fungal growth develop on the underside of the leaves, which become blotched on the upper surface. Affected seedlings are severely checked and lettuces may later be attacked by gray mold (see below). These mildews are most troublesome on overcrowded seedlings growing in very humid conditions. Prevent the diseases by sowing seed thinly in sterilized, well drained soil or compost, and ventilate carefully to reduce humidity. Should mildew occur, remove diseased leaves and spray with mancozeb or zineb. If brassica seedlings are affected, the mildew may also be controlled by applying either of the fungicides chlorothalonil or captafol.

Stems collapsing

Damping off is usually due to species of the soil- and water-borne fungi *Phytophthora* and *Pythium*. Seedlings of lettuce, tomato, mustard and cress are most susceptible to infection, and collapse at ground level. Overcrowding encourages the disease, therefore sow thinly and use sterilized soil or compost of a good tilth. The organisms that cause damping off are often present in unsterilized compost and soil, particularly if it is compacted causing poor aeration. Over-watering can also induce damping off. Use clean water to prevent infection by water-borne organ-

isms which build up in dirty tanks and butts. Give adequate light but not too much heat. Check slight attacks by watering with captan, zineb or Cheshunt compound after removal of the dead seedlings.

Wirestem fungus, caused by *Rhizoctonia solani*, is a disease of brassica seedlings, particularly cauliflowers, but the same fungus can also affect seedlings of other vegetables that are being raised under glass or polyethylene. Stems of affected brassica seedlings shrink at ground level before they topple, but other seedlings damp off as described above, due to blackening and death of the roots. Lettuce seedlings affected by this fungus usually succumb to gray mold (see below) fairly soon afterwards so that the original cause may be overlooked. Prevent this disease by sowing thinly in a good tilth and avoid over-watering. Use sterilized compost to help prevent infection. The fungus can be controlled by dicloran and quintozene. Both chemicals should be raked into the soil before sowing seed where this disease is known to be troublesome.

DISEASES OF MATURE CROPS

The diseases described below may, unless otherwise stated, affect any crop being grown in the greenhouse or under cover. Vines and peaches are treated separately at the end of this section.

Leaves discolored

Faulty root action is due to over- or under-watering or poor transplanting and can cause irregular yellow or brown blotches on the leaves. Prevent this by careful planting and correct cultural treatment. Applications of a foliar feed should help to overcome the trouble, but in severe cases it may be necessary to mound sterile compost around the base of the stem into which new roots can grow as the plant recovers.

Magnesium deficiency is common on tomatoes and eggplants. Orange-yellow bands develop between the veins on the lower leaves, which gradually turn brown as the symptoms spread progressively upwards. Spray at the first signs of trouble with $\frac{1}{2}$ lb magnesium sulfate in 3 gal of water, to which is added a spreader. Repeat at seven

to ten day intervals. Affected plants still produce good crops.

Leaves moldy

Tomato leaf mold (*Cladosporium fulvum*) affects only tomatoes grown under glass or polyethylene. A purple-brown mold develops on the lower surface of leaves which show yellow blotches on the upper surface. These symptoms may be overlooked as affected leaves are subsequently often attacked by gray mold. Grow resistant varieties and keep the greenhouse temperature less than 21°C/70°F. Ventilate well since the disease is encouraged by humid atmospheres. At the first signs of trouble spray with benomyl or mancozeb, or apply Exotherm Termil every seven days until symptoms disappear.

Leaves and stems rotting

Gray mold (*Botrytis cinerea*) is a common problem under glass, affecting particularly grapes, strawberries, cucumbers and tomatoes. Lettuce tends to wilt due to attack at ground level. Affected stems, fruits and leaves rot and become covered with a gray-brown velvety fungus growth. Sometimes the fungus does not rot tomato fruits but produces pinpoint spots, each with a pale green ring, known as water spots, which can still be seen on ripe fruit. Spores of the fungus infect plants through wounds and dead and dying tissues, or by contact between diseased and healthy tissues. Remove dead leaves and over-ripe fruits promptly to avoid infection. Ventilate greenhouses carefully to reduce humidity and water early in the morning, not at night. Over-wintering plants should be sprayed with thiram every three or four weeks. Prevent infection of grapes and strawberries by spraying with benomyl or thiophanate-methyl, as the first flowers open, repeating twice at ten day to two week intervals, or with captan or thiram except on fruit to be preserved. Fumigate an affected greenhouse with Isotherm Termil bombs.

Stems wilting

Foot and root rot can be due to various fungi, including *Thielaviopsis basicola* and species of *Fusarium*, as well as those fungi which cause damping off and wirestem of seedlings

(see above). The top growth wilts or collapses completely because these soil- and water-borne organisms attack the roots and stem bases. Prevent this by the use of clean water and by changing or sterilizing the soil at least once every three years, or by the use of sterile compost. Plant carefully, and tease out roots of pot-bound plants. Do not over- or under-water as plants suffering from faulty root action (see above) are very susceptible to attack. If foot rot occurs, water with a solution of captan, Cheshunt compound or zineb, or dust at the base of the plant with dry bordeaux powder. When tomatoes are affected, place fresh sterilized compost around the base of the stems and spray all plants with a foliar feed to encourage the development of new roots.

Verticillium wilt is caused by species of the fungus *Verticillium*. The larger leaves wilt during the day, particularly on hot days, but recover at night. Affected plants may lose their older leaves. Brown streaks are seen running lengthways in the tissues if the base of the stem is cut longitudinally. Destroy badly affected plants. On less severely diseased plants, encourage the development of new roots as recommended above for foot and root rot. Prevent the disease by using sterilized soil or compost. Alternatively, grow only those varieties that are certified to be verticillium and fusarium resistant.

Tomato stem rot (*Ascochyta lycopersici*) causes a sudden wilting of mature plants. A brown or black canker develops on the

Protected crops 2

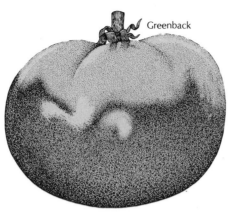

Blotchy ripening

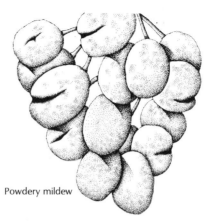

Greenback

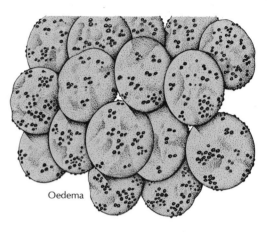

Powdery mildew

Oedema

stem, usually at ground level, and small black specks, which are the fruiting bodies of the fungus, can just be seen with the naked eye all over the diseased tissues. These produce many spores which over-winter and act as a source of infection the following season. It is essential, therefore, to burn all debris and to sterilize the greenhouse and equipment at the end of the season if this disease has occurred. Destroy badly affected plants and spray the stem bases of the rest of the crop with benomyl or captan. Less severely diseased plants may be saved by cutting out affected tissues and applying a paste of captan mixed with a little water, or by painting them with a solution of benomyl.

Flowers dropping

Tomato flower drop is almost always due to dry conditions at the roots. The flowers may open, but break off from the stalk at the joint and fall to the ground. Prevent this trouble by adequate but careful watering.

Fruits failing to develop normally

Withering of young cucumbers starting at the blossom end is due to uneven growth resulting from irregular watering. Remove all the fruits from an affected plant to rest it, and spray the foliage with a foliar feed if it has a poor color. Later-developing fruits should be normal once the plant regains its vigor, providing there is no root disease present. Prevent further trouble by watering cucumbers carefully and regularly.

Chats (small tomato fruits) may form on plants which are dry at the root, but poor pollination caused by cold nights and a dry atmosphere may also be responsible. Encourage pollination by syringing the foliage in the morning and again during the day when the weather is hot.

Dry set of tomatoes is also due to poor pollination. It is caused by the atmosphere being too hot and too dry. The fruits remain $\frac{1}{8}$ in across and become dry and brown. Syringe the foliage as described for chats.

Fruits discolored

Blossom end rot of tomatoes shows as a circular and depressed brown or green-black patch on the skin at the blossom end of the fruit (the end farthest away from the stalk). In most cases it is due to a shortage of water at a critical stage in the development of young fruit. Prevent this by seeing that the soil is never allowed to dry out completely. All the fruit on one truss may be affected but those developing later should be normal if the plant has a good root system and is looked after carefully.

Greenback and blotchy ripening of tomatoes show as hard green or yellow patches on the fruits. The former occurs on the shoulder of the fruit and the latter on any part. Both may be encouraged by high temperatures and a shortage of potash; greenback is also caused by exposure of the shoulder to strong sunlight, and blotchy ripening may occur where nitrogen is deficient. Prevent these troubles

by adequate and early ventilation, by ensuring that plants have sufficient shade, and by correct feeding and watering. Grow tomato varieties resistant to greenback, such as 'Eurocross A', 'Eurocross BB', 'Amberley Cross', 'Craigella' and 'Alicante'.

Bronzing of tomatoes is caused by tobacco mosaic virus. Brown patches develop beneath the surface, usually at the stalk end, and give a bronzed patchy appearance to the young fruit. When cut open the patches show as a ring of small dark spots beneath the skin. With severe infection depressed streaks which fail to ripen may radiate from the stalk end. The internal tissues of such fruits show large brown corky areas. Plants bearing bronzed tomatoes would have shown other symptoms such as stunted growth or mottled foliage earlier in the season and should have been destroyed when these symptoms first appeared. (See Virus diseases under Outdoor tomatoes, page 51.)

Fruits rotting

Gray mold (*Botrytis cinerea*) can attack various crops. For details, see previous page.

Fruits bitter

Bitter cucumbers can be due to an excess of nitrogen in the soil or irregular growth. Avoid excessive use of nitrogenous fertilizers, and maintain even growth by watering carefully. Since pollination of the fruit can also result in bitterness, grow varieties having mostly female flowers, for example, 'Pepinex'.

VINES

The most serious disorder to affect vines grown under glass is powdery mildew.

Leaves, shoots and fruits with fungal growth
Powdery mildew (*Uncinula necator*) shows a soft white floury coating of fungus spores on the leaves, young shoots and fruits. Affected berries drop if attacked early, but in later attacks become hard, distorted and split, and are then affected by secondary fungi such as gray mold. Ventilate carefully since the disease is encouraged by humidity. Avoid overcrowding the shoots and leaves and provide some temporary heat in a cold house. Avoid also dryness at the roots. At the first sign of mildew spray or fumigate with dinocap, spray or dust with sulfur, or spray with benomyl or thiophanate-methyl. Up to four applications may be needed, whichever chemical is used. In winter, after removing the loose bark paint the vine stems with a solution of sulfur made up as follows: mix equal parts of flowers of sulfur and soft soap to form lumps the size of golf balls. Put one lump into a jam jar with a little water and stir well with the brush used to paint the stems.

Leaves discolored

Scorch is due to the sun's rays striking through glass on to moist tissues on a hot day. It shows as large brown patches which soon dry out and become crisp. Prevent this by careful ventilation to reduce humidity and remove affected leaves.

Protected crops 3

Magnesium deficiency shows as a yellow-orange discoloration between the veins, but in some varieties the blotches may be purple. Later the affected areas turn brown. Spray with $\frac{1}{2}$ lb of magnesium sulfate in 3 gal of water plus a spreader such as soft soap or a few drops of mild washing-up liquid. Repeat applications once or twice at two week intervals.

Leaves with small globules
Exudation of small round green or colorless droplets from the leaves is quite natural and usually goes unnoticed. However, in the spring the transparent globules may become very noticeable on the young foliage. The symptoms are most obvious on plants growing in a very humid atmosphere and they indicate that the root action is vigorous and the plant is in good health. Nevertheless, ventilate carefully to reduce the humidity and prevent other troubles.

Vine dying
Honey fungus (*Armillaria mellea*) frequently kills indoor and outdoor vines. It attacks via the soil, causing white fan-shaped growths of fungus to develop beneath the bark of the roots and the main stems at and just above ground level. Dark brown root-like structures known as rhizomorphs develop on the affected tissues, grow out through the soil and spread the disease. Dig out dead and dying plants together with as many roots as possible. If the greenhouse is vacant, sterilize the soil with 2 per cent formalin. Alternatively, change the soil completely before replanting with fresh stock.

Fruit failing to develop normally
Shanking is due to one or more unsuitable cultural conditions. The stalk of the berry shrivels gradually until completely girdled. Odd berries or small groups of berries then fail to color and develop naturally at the early ripening stage. The berries are watery and sour and those of black varieties turn red while those of white varieties remain translucent. Ensure that over- or under-watering or stagnant soil conditions are not responsible. Reduce the crop for a year or two until the vine has regained its vigor.

When shanking occurs early in the season, cut out the withered berries and spray the foliage with a foliar feed.
Splitting of berries most commonly occurs as a result of powdery mildew (see above). However, it is sometimes due to irregular watering. Remove affected berries before they are attacked by secondary organisms such as gray mold, and water before the soil dries out.
Scald is caused by the sun's rays striking through glass on to moist tissues on a hot day. Ventilate carefully to reduce the humidity. Remove affected berries showing sunken discolored patches.
Oedema occurs when the roots of an affected plant take up more water than the leaves can transpire and is due to extremely moist conditions in the soil, the atmosphere, or both. It shows as small warts or pimples on the stalks and sometimes on the berries and even on the lower leaf surface. These outgrowths may break open and then have a blister-like or white powdery appearance, or they may become rusty-colored and show as brown scaly patches. Do not remove the affected parts as this will make matters worse. Maintain drier conditions both in the air and soil; with correct cultural treatment the affected plant should eventually recover.

PEACHES
The following remarks on split stone also apply to nectarines.

Fruit failing to develop normally
Split stone shows as a cracking of the fruit at the stalk end, forming a hole large enough for the entry of earwigs. The stone of such a fruit is split and the kernel is either rotting or absent. Affected fruits are susceptible to secondary rotting. This trouble can be due to the soil being too acid. Lime to bring the pH up to 6.7–7.0. Poor pollination can also cause split stone, therefore hand-pollinate flowers by passing cotton-wool or a soft camel hair brush from flower to flower. The commonest cause of this trouble, however, is an irregular water supply. Prevent this by watering in dry periods and mulching to conserve moisture. In particular, ensure that the soil is never allowed to dry out.

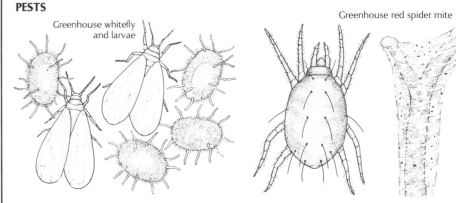

PESTS
Greenhouse whitefly and larvae

Greenhouse red spider mite

Greenhouse red spider mites (*Tetranychus urticae*) are tiny, eight-legged creatures that can occur in large numbers on the undersides of leaves. They are just visible to the naked eye but a hand lens is necessary to see them clearly. Despite their common name, these mites are yellow-green with black markings; they only become orange-red in the fall when they hibernate. Their sap feeding causes the upper surface of the leaves to become discolored by a fine mottling. In severe infestations leaves dry up and the plants become festooned with a silken webbing produced by the mites. Maintaining a damp atmosphere helps to check this pest but insecticides such as malathion, dimethoate or formothion will also be needed at seven day intervals until the pest has been controlled. Take care when applying these chemicals to cucumbers and melons as they may be damaged by insecticides. Avoid this risk by spraying in the evening when temperatures are cooler, and by making sure the plants are not dry at the roots. As an alternative to insecticides this pest can be controlled by introducing a predatory mite, *Phytoseiulus persimilis*. For details see page 3.
Peach-potato aphid (*Myzus persicae*) is usually the most troublesome greenfly in greenhouses. These sap-feeding insects can be found on the underside of the leaves and on the stems and shoot tips. Infested plants tend to grow slowly with puckered leaves which are often shiny because of a sticky coating of honeydew that the aphids excrete. Sooty molds grow on honeydew under humid conditions, blackening the leaves and fruits. As aphids grow they shed their skins. These white skins become stuck on the honeydew and can be seen more easily than the insects themselves. In addition to direct damage, aphids can also transmit certain viruses. Suitable insecticides include pirimicarb, pirimiphos-methyl or a pyrethroid compound. Use the last-mentioned insecticide if the crops are ready for eating.
Greenhouse whitefly (*Trialeurodes vaporariorum*) is a major pest of greenhouse plants. Both the small, white, moth-like adults and their flat, oval, white-green, scale-like larvae feed by sucking sap from the underside of leaves. Like aphids, adults and larvae excrete honeydew which allows the growth of sooty mold. Early treatment with pirimiphos-methyl or a pyrethroid compound such as pyrethrum, resmethrin or bioresmethrin will prevent damage occurring. Spray heavy infestations several times at three to four day intervals. Greenhouse whitefly can be controlled during the summer by introducing a parasitic wasp, *Encarsia formosa*, (see page 3). Alternatively, coat yellow-painted surfaces with vaseline or tanglefoot. This will attract and trap the flying adults, and gives good control.

Other vegetables

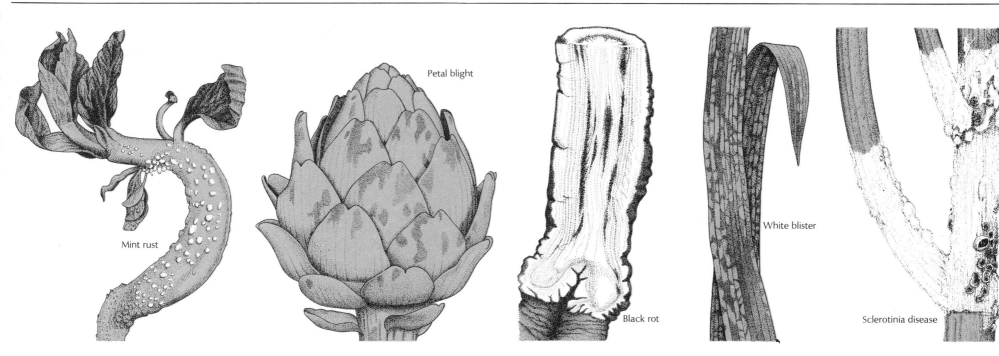

Petal blight

Mint rust

Black rot

White blister

Sclerotinia disease

Globe artichoke—heads blotched

Petal blight (*Itersonilia perplexans*) is a localized disease but, where it occurs, can be troublesome in a wet season. The fungus affects the heads before they mature, causing pale brown circular spots to develop followed rapidly by complete browning and even rotting of the heads. In gardens where this disease is known to occur spray with zineb soon after the buds begin to develop and repeat at two week intervals until about three weeks before harvesting. Remove and burn diseased heads. Do not grow chrysanthemums or dahlias anywhere near globe artichokes as the disease may spread via these ornamentals.

Globe artichoke—roots with pests visible

Root aphids (*Trama troglodytes*) suck sap from the roots causing poor growth and, in severe cases, wilting. The aphids are soil-colored but their presence is made obvious by the white waxy powder that they secrete over the roots and soil particles. Control these aphids by drenching the root area with a solution of either malathion or a systemic insecticide.

Jerusalem artichoke—stems and tubers rotting

Sclerotinia disease (*Sclerotinia sclerotiorum*) can attack the tubers during storage but is most troublesome during the growing period when it attacks the stems. These become rotten but do not always show the typical white fluffy growth of the fungus on the outer tissues. More often the fungus develops within the stems where it produces numerous large ($\frac{1}{2}$ in or more) irregular-shaped resting bodies which are only seen if the stem is broken. It is essential, therefore, to keep a careful watch on Jerusalem artichokes during the growing season, particularly if this disease has occurred in previous years or on other crops. Any plant showing discolored patches or rotting of the stems should be removed immediately and burned before the resting bodies fall into the soil and contaminate it. Grow Jerusalem artichokes on a fresh site the following year.

Mint—leaves and stems with fungal growth

Mint rust (*Puccinia menthae*) first occurs in the spring as swollen distorted shoots covered with small orange spore pustules. Remove and burn affected shoots to prevent the disease from spreading. If left the fungus spreads to the leaves which show pale spots on the upper surface, dry up and fall off. In spite of these symptoms it is often possible to harvest enough mint for use in cooking. However, where the bed is severely infected cover it in the fall with dry straw and set it alight on a windless day to burn off the top growth. To establish a new healthy bed using runners from an infected area, first wash the cuttings in cold water, then place them in hot water at 40°–45°C/105°–115°F for ten minutes. Plunge them into cold water again before planting out.

Salsify and scorzonera—leaves blistered

White blister (*Albugo tragopogonis*) shows as glistening white pustules of fungus spores on the leaves. Growth may be stunted and root development poor. It is only likely to be troublesome on overcrowded plants so good spacing should check the disease. Cut off and burn affected leaves and do not compost them or dig them into the soil since they will contain the resting bodies of the fungus. Grow crops on a fresh site the following year.

Seakale—blackening of the tissues

Black rot (*Xanthomonas campestris*) is a bacterial disease that causes the leaves to turn yellow and the veins black. A black dotted ring shows on cut surfaces of the leaf stalks, stems and on the roots, particularly when the side roots (thongs) are removed for cuttings. Destroy affected plants and do not use crowns for forcing if they show black streaks. The disease occurs most commonly in warm wet summers and is most serious where the drainage is faulty. Check the disease by carrying out a strict rotation of crops and, if possible, by improving the drainage of the site.

Seakale—roots distorted

Club root (*Plasmodiophora brassicae*) can be troublesome on seakale. The roots become swollen and distorted and do not function properly. This causes the foliage to become discolored and the leaves to wilt in hot weather. Burn diseased debris and grow seakale on a fresh site where no other crucifer has been grown, in well drained soil which has been limed. Use quintozene in the planting furrows or transplant solution.

Protected ornamentals 1

Introduction

Ornamentals are plants that are grown for their decorative value. The troubles dealt with in this section are, therefore, primarily those that spoil the appearance of a plant. Ornamentals are also susceptible to certain physiological disorders that are not mentioned specifically in this section, and it is advisable always to read the section on physiological disorders (pages 5–7) when trying to identify the cause of the trouble affecting any unhealthy ornamental. Such disorders should not arise if ornamentals are sown or planted correctly, in well drained soil of the right pH, and given the correct cultural treatment. However, good health cannot be achieved unless high-quality seed or plants are purchased. Do not buy cheap lots of plants left over at the end of the selling period without first checking that they are in good condition, otherwise pests and diseases may be introduced into a garden, particularly in the case of bulbous plants. The permanence of perennial herbaceous plants, trees and shrubs does favor the build-up of pests and diseases; but this can be prevented by good hygiene and by observing the principles listed in the Introduction on pages 2–3.

BULBOUS PLANTS

This section treats problems that are specific to plants having bulbs, corms, tubers or rhizomes.

Plant stunted

Non-rooting of hyacinth bulbs is a physiological disorder, the precise cause of which is not known. The leaves do not develop at the normal rate and the inflorescence remains stunted. The roots of an affected bulb are either lacking or poorly developed. This problem can be caused by the temperature being too high during storage or forcing, or by forcing or lifting too early. Unfortunately it is not possible to detect in advance those bulbs in which the non-rooting tendency has developed.

Plant wilting

Bacterial wilt (*Xanthomonas begoniae*) causes wilting and spotting on leaves of winter-flowering begonia hybrids derived

from *B. socotrana* and *B. dregei*. Burn severely diseased plants and do not propagate from them. If they are only slightly diseased cut out affected parts and decrease the temperature and humidity of the greenhouse. This will reduce the spread and severity of the disease, but it will also delay flowering. Disinfect the greenhouse after a severe attack of the disease.

Leaves discolored

Leaf scorch (*Stagonospora curtisii*) causes brown blotches to appear on the leaves of hippeastrum (amaryllis), particularly at the leaf bases, and also on the flower stalks and petals. The affected tissues usually rot and become slimy. Cut out such tissues and burn them. Spray or dust affected plants with sulfur or zineb.

Unsuitable cultural conditions can check the growth of hippeastrums, causing red blotches or streaks (or both) to appear on the leaves, flower stalks and bulbs. This trouble is usually caused by over- or under-watering or malnutrition; prevent it by maintaining even growth through good cultural treatment; see the Introduction, pages 2–3.

Leaves distorted

Tarsonemid mites are a group of tiny creatures that infest the growing points of certain greenhouse plants. The bulb scale mite (*Steneotarsonemus laticeps*) lives in the neck of narcissus and hippeastrum bulbs. They cause a distinctive sickle-shaped curvature of the leaves and a saw-toothed notching along the margins. The flower stems become stunted and distorted, again with a saw-toothed scar along the edges of the stem. The cyclamen mite (*Tarsonemus pallidus*) and broad mite (*Polyphagotarsonemus latus*) live inside the leaf and flower buds of plants such as cyclamen, *Hedera* (ivy), begonia, impatiens, saintpaulia and *Sinningia* (gloxinia). Their feeding causes stems and leaves to become scarred and frequently to be distorted into spoon-like shapes. The growing points may be killed and the flowers are either distorted or fail to develop. There are no chemicals available to amateur gardeners for controlling these mites. Burn all infested plants.

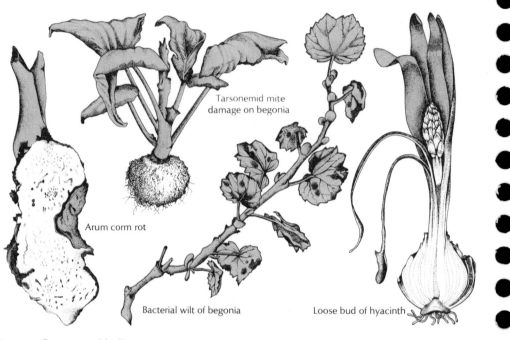

Tarsonemid mite damage on begonia

Arum corm rot

Bacterial wilt of begonia

Loose bud of hyacinth

Leaves, flowers and bulbs rotting

Soft rot (*Erwinia carotovora*) causes hyacinths to develop a soft, slimy, evil-smelling rot of the leaves and bulbs. It often commences in the inflorescences when florets have withered through a physiological disorder known as blindness; for details see page 59 under Buds withering. If the rot has not advanced too far it may be possible to save the bulbs for planting outside by cutting out all infected tissue.

Roots or tubers eaten

Black vine weevil grubs (*Otiorhynchus sulcatus*) are plump white maggots about $\frac{1}{2}$ in long with light brown heads. Plants grown from tubers are particularly susceptible but many other plants may be attacked. Usually the first symptom that is noticed is the plant wilting and, when it is tipped out of its pot, most of the roots are seen to have been destroyed. Once these weevil maggots have attacked a particular plant it is not likely to recover. Susceptible plants can be given some protection by adding chlorpyrifos granules or naphthalene flakes to the compost when potting up.

Bulbs, corms or tubers rotting

Basal rot may be caused by various fungi, and affects mainly *Lilium* and *Lachenalia*. The roots and base of the bulb rot, resulting in stunting of the top growth and discoloration of the leaves. Discard badly affected bulbs. In less severe cases cut out diseased roots and tissues, or scales in the case of lily bulbs. Then dip the bulbs in a solution of captan or benomyl before re-potting. Prevent such troubles by using only sterile compost and clean pots.

Begonia tuber rot and cyclamen corm rot usually occur as a result of frost damage during storage. The tissues become soft and have a sweetish smell. Prevent these rots by ensuring that tubers and corms of the respective plants are stored carefully in a frost-proof place.

Arum corm rot (*Erwinia carotovora*) can cause serious damage wherever arums (*Zantedeschia* spp and hybrids) are grown under glass in large numbers. The plants wither and collapse due to rotting of the corms; these may develop extensive brown areas with rotting roots arising from them. The corm lesions can lie dormant during storage but

Protected ornamentals 2

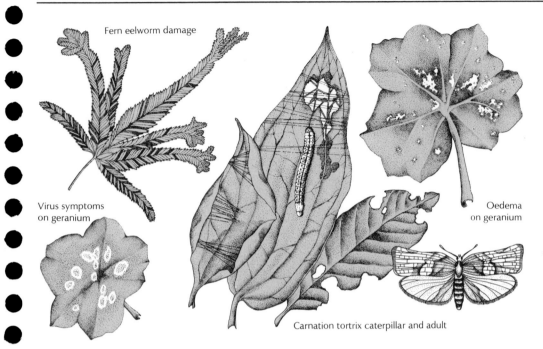

Fern eelworm damage

Virus symptoms on geranium

Oedema on geranium

Carnation tortrix caterpillar and adult

when the corms are replanted the rot progresses rapidly. Destroy badly infected plants and disinfect the house. Sterilize the soil where diseased plants have been growing in beds. Examine corms when removing them from store and cut out any brown areas. Then steep them for two hours in a 2 per cent formalin solution as a safety measure before planting them out.

Inflorescence loose
Loose bud of hyacinth, in which the stem below the flower bud fractures completely at an early stage of growth, is usually caused by storing bulbs at too low a temperature. Bulbs that have been moved from the cold store into a very warm place are particularly susceptible. Loose bud may also be caused by incorrect lifting or forcing. Unfortunately it is impossible to detect the tendency for loose bud in a consignment of bulbs.

Buds withering
Blindness of bulbous plants is usually caused by the soil being too dry at a critical stage of growth. Prevent this by making sure that the compost never dries out. Less frequently it is

caused by storing bulbs before planting in conditions that are too hot and dry. Prevent this either by potting up immediately on obtaining bulbs, or by storing them in the proper conditions. The flower buds of affected bulbs turn brown and wither at an early stage. Such bulbs can be planted out in the garden but will not flower for a year or two.

GENERAL PLANTS
The pests and diseases mentioned in this section may, unless otherwise stated, affect any type of pot plant, including those with bulbs, corms, tubers or rhizomes. For pests and diseases of seedlings, see Plants raised from seed, pages 68–70.

Leaves eaten
Carnation tortrix caterpillars (*Cacoecimorpha pronubana*) feed on a very wide range of plants and can be found throughout the year in heated greenhouses. The caterpillars grow up to $\frac{3}{4}$ in long and are pale green with brown heads. They fold over the edge of a leaf with silken threads, or bind two leaves together, and when small feed unnoticed by grazing away the inner surfaces of these leaves. Later

these caterpillars eat holes in the foliage. Control light infestations by searching for and squeezing the caterpillars' hiding places. Otherwise spray the plants thoroughly with pirimiphos-methyl or trichlorphon when signs of damage are seen. Other caterpillars that can be found on greenhouse plants include those of the angle shades moth (*Phlogophora meticulosa*) and the silver-Y moth (*Autographa gamma*). These feed in the open on the foliage and flowers but may be difficult to find since they are active mainly at night. Control these pests by hand-picking or by applying the above insecticides.

Slugs (various species) can damage most plants, especially during the early stages of growth. They frequently leave a slime trail on the foliage, which distinguishes their damage from that caused by caterpillars. Control them by scattering slug pellets based on either methiocarb or metaldehyde on to the soil surface.

Leaves discolored
Faulty root action may be caused by over- or under-watering, malnutrition or poor potting. It results in irregular yellow or brown blotches on the leaves, or complete discoloration of the foliage, and premature leaf-fall. Prevent such troubles by careful potting up and correct cultural treatment for the type of compost being used. Applications of a foliar feed should help to overcome the troubles but in severe cases it may be necessary to re-pot the affected plant.

Tip scorch of the leaves of house plants such as aspidistra, chlorophytum and sansevieria may be caused by the air being too hot or dry, or by faulty root action (see above). Affected plants should recover once the scorched leaves have been removed and the correct cultural treatment given. In the case of saintpaulia, anthurium and palms such as kentia it may be necessary to place the pot in a larger container packed with damp moss or peat in order to create a humid atmosphere.

Sun scorch of leaves usually shows as pale brown blotches (often elliptical) across the foliage. It is caused by the sun's rays on a hot day passing either through glass on to moist foliage, or through a flaw in the glass which acts as a lens to intensify the rays. Prevent

scorch in greenhouses by careful ventilation to reduce humidity. For house plants, move them away from the window or turn them round each day. Do not stand pot plants in front of windows containing "bottled" or faceted glass, since they also intensify the sun's rays. Remove affected leaves.

Leaf spots are caused by a variety of fungi. In practically all cases they produce brown or black spots on the leaves, but on some hosts the spots have a purple border or they may have pinpoint-sized black dots scattered over them. Remove affected leaves and spray with mancozeb or zineb. If further trouble occurs the plants may be lacking in vigor because of faulty root action, in which case see above.

"Ring pattern" on saintpaulias and achimenes is caused by a sudden chilling of the leaves from watering overhead in sunlight. Affected leaves develop large yellow rings. Prevent this by careful watering.

Viruses such as tomato spotted wilt and cucumber mosaic affect a wide range of plants. In general the symptoms are mottled, blotched or striped leaves, affected parts being pale green, yellow or black. The leaves may also be distorted and the plants

---

SOOTY MOLD

Some sap-feeding insects such as aphids, whiteflies, scales and mealybugs excrete a sugary liquid known as honeydew. Since these insects feed mainly on the undersides of leaves the honeydew drops down on to the upper surfaces of leaves growing below the actual infestation. Such leaves become sticky, and under damp conditions various black, non-parasitic fungi known as sooty mold rapidly develop. They do not directly harm plants because they grow on the honeydew, although the amount of light and air reaching the foliage is reduced. Remove sooty mold by wiping the leaves with a soft, damp cloth. Good ventilation makes the atmosphere drier and thus less suitable for the growth of sooty mold, but the best cure is to identify and control the pest that is producing the honeydew.

Protected ornamentals 3

stunted. Destroy any plant showing these symptoms. A valuable plant such as an orchid may be kept but it will always produce discolored leaves and the trouble may spread to previously healthy plants.

Greenhouse thrips (*Heliothrips haemorrhoidalis*) are thin yellow or dark brown insects about $\frac{1}{10}$ in long that live mainly on the upper surfaces of leaves and on flowers. They feed by sucking sap and cause a dull green or silvery discoloration of the foliage, which is also marked by minute black spots caused by the thrips' excretions. Control this pest by spraying thoroughly with a pyrethroid compound, derris, malathion or a systemic insecticide.

Leaves with corky patches

Oedema, or dropsy, is caused by the atmosphere being too moist or the soil too wet. It shows as pale pimple-like outgrowths on the undersurfaces of the leaves and on the stems. The outgrowths later burst and then become brown and powdery or corky. The most susceptible plants are eucalyptus, ivy-leaved pelargonium, peperomia and camellias—the last mentioned develops large scabby patches on the undersurfaces. Improve the cultural conditions by careful watering and by ventilating the greenhouse. Do not remove affected leaves since this will only make matters worse.

Corky scab of cacti is caused either by a lack of light and the humidity being too high, or by over-exposure to sunlight. It occurs most frequently on *Epiphyllum* and *Opuntia* and shows as irregular rusty or corky spots which develop into sunken patches as the tissues beneath die. Where the trouble is very unsightly propagate from the affected plant and ensure that new plants are given correct cultural treatment and are not exposed to too much light.

Leaves blotched

Chrysanthemum eelworm and fern eelworm (*Aphelenchoides ritzema-bosi* and *A. fragariae*) are microscopic worm-like animals that live inside leaves. Many different plants may be infected, although in greenhouses the main hosts are those indicated by the pests' common names. Infested parts of the leaves turn brown. At first these areas are clearly separated by the larger leaf veins from the green, healthy parts, but eventually the brown areas coalesce and the whole leaf dies. None of the chemicals available to amateur gardeners controls eelworms, and infested plants should be burned. However, it is possible to give chrysanthemum stools a hot water treatment so that they will subsequently produce cuttings free of eelworms. Wash the dormant stools free of all soil and then plunge them in hot water at 46°C/115°F for five minutes—it is important that the time and temperature are exact. After treatment plunge the stools into cold water.

Leaves mined

Chrysanthemum leaf miner grubs (*Phytomyza syngenesiae*) tunnel the leaves of chrysanthemum and related plants such as cineraria (*Senecio cruentus* hybrids) and gerbera. These mines show on the leaves as whitish-brown lines meandering through the leaf and, in heavy infestations, leaves may lose almost all their green color. A single application of pirimiphos-methyl controls this pest if applied as soon as mining begins, but if the plants are badly infested three applications of insecticide at ten day intervals will be necessary.

Leaves with visible fungal growth

Powdery mildews are common on chrysanthemums, begonias and cinerarias, and occur occasionally on other plants. The symptoms are white powdery spots on the leaves and sometimes the stems. Ventilate the greenhouse well since the fungi are encouraged by a humid atmosphere. Plants that are dry at the roots are more susceptible to infection, so water before the compost dries out completely. Fumigate the greenhouse with dinocap smokes or spray with dinocap or benomyl. Remove severely affected leaves.

Rusts can affect chrysanthemums, fuchsias, pelargoniums, cinerarias and carnations. On fuchsias and cinerarias orange powdery pustules develop on the leaves, predominantly on the lower surfaces. On other plants the pustules produce masses of chocolate-colored spores. Remove and burn affected leaves. If severely infected, destroy the whole

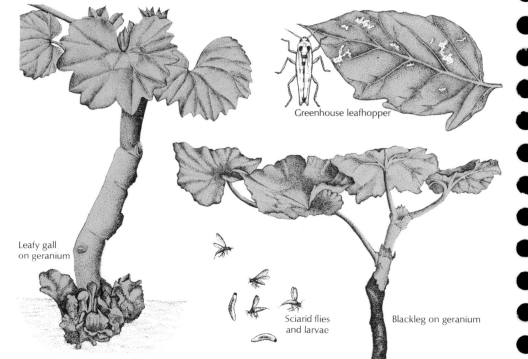

Greenhouse leafhopper

Leafy gall on geranium

Sciarid flies and larvae

Blackleg on geranium

plant. Ventilate the greenhouse well to reduce the humidity of the atmosphere, and make sure when watering that droplets do not remain on the leaves. Spray at seven to ten day intervals with zineb or mancozeb.

Leaves with pests visible

Greenhouse whitefly (*Trialeurodes vaporariorum*) is one of the most common and troublesome of greenhouse pests. The adults are small white moth-like insects that occur in large numbers on the undersides of leaves. They readily fly up from the plants when disturbed. Their whitish-green larvae and pupae resemble scales and, like the adults, feed by sucking sap from the undersides of leaves. The immature stages are relatively immune to insecticides but the adults can be killed by spraying with a pyrethroid compound or pirimiphos-methyl. Alternatively in the greenhouse, adult whiteflies can be trapped on the surface of yellow-painted boards by coating the surface with either vaseline or tanglefoot, since whiteflies are attracted to these chemicals. During the summer another means of control is to introduce a parasitic wasp called *Encarsia formosa*. For further details, see page 3.

Peach-potato aphid and mottled arum aphid (*Myzus persicae* and *Aulacorthum circumflexum*) are both species of greenfly that suck sap from a wide range of plants. The former is either pink or yellow-green, both types often occurring together on the same plant, while the latter is yellow-green with a dark horseshoe marking on its back. As the aphids grow they shed their skins, which become stuck on the leaf surface where they are held by the sticky honeydew that aphids excrete. These skins are white and sometimes mistaken for whitefly or some other pest. Control aphids by applying pirimicarb, pirimiphos-methyl, pyrethroid compounds or malathion.

Soft scales (*Coccus hesperidum*) are sap-feeding insects that live on the stems and undersides of leaves near the main veins. For a description of these insects and their control, see page 61 under Scale insects.

Protected ornamentals 4

Leaves mottled

Greenhouse red spider mites (*Tetranychus urticae*) attack most greenhouse and house plants. The mites are only just visible to the naked eye and can be found mainly on the undersides of leaves where they feed by sucking sap. The early symptom of damage is a fine speckling of the upper surfaces of affected leaves. As the infestation develops the foliage becomes yellow or a dull pale green and the older leaves start to dry up and fall. At this stage the mites produce a fine silken webbing between the leaves, and large numbers of them may swarm over the highest parts of the plant. Spray thoroughly when symptoms are first seen with malathion, dimethoate or formothion, with at least two further treatments at seven day intervals. Alternatively, control them by introducing the predatory mite *Phytoseiulus persimilis*; for further details, see page 3. Red spider mites thrive under hot dry conditions, so maintaining a humid atmosphere may help to deter this pest.

Greenhouse leafhoppers (*Zygina pallidifrons*) suck sap from the undersides of leaves and cause white, pinhead-sized dots to appear on the upper surfaces. In heavy attacks these dots coalesce and most of the leaves' green color is lost. Adult leafhoppers are about $\frac{5}{8}$ in long and pale yellow with two V-shaped gray markings on their back. The nymphal stages are creamy-white. As they grow they periodically shed their skins, which remain attached to the undersides of the leaves. Control them by spraying with any of the insecticides malathion, pirimiphos-methyl, methoxychlor or a pyrethroid compound such as resmethrin.

Stems or crowns rotting

Blackleg (various organisms) affects pelargonium cuttings and sometimes the mature plant. The stem bases become soft, black and rotten, and affected plants die. Prevent this disease by using sterile compost and pots, and by hygienic cultural conditions, including the use of clean water. Destroy severely diseased cuttings, but in the case of valuable plants it may be possible to propagate by taking a fresh cutting from the top of a diseased plant.

Foot, crown and root rot may be caused by black root rot fungus or other soil or water-borne fungi. These organisms cause a brown or black rot of the tissues at the base of the stems, around the crowns or at the roots, and the top growth wilts or collapses. Prevent these diseases by using sterilized compost and pots, and by using clean water. Pot up carefully and tease out the roots of pot-bound plants. Control by watering with a solution of benomyl combined with ethazol. Alternatively, water the soil thoroughly with a solution of Banrot. In severe cases re-pot, using a smaller pot if necessary, in sterile compost after having removed all dead parts including roots. Spray developing leaves with a foliar feed.

Gray mold (*Botrytis cinerea*) causes plants to decay and affected leaves and flowers to become covered with a gray-brown mass of fungal spores. The petals may also develop numerous small red or brown spots. Gray mold spores are always present in the air and infect plants through wounds and dead or dying tissue. Infections can also occur between diseased and healthy tissues. Prevent gray mold by good hygiene and by removing dead leaves and flowers promptly. Ventilate the greenhouse carefully to reduce humidity, and water early in the morning and not at night. Once the disease has appeared on any type of plant, spray with benomyl or thiophanate-methyl. Alternatively, fumigate with Isotherm Termil bombs. In the case of cyclamen affected by gray mold around the crown, dust with captan.

Carnation wilt is caused by the fungi *Verticillium albo-atrum* and *Fusarium oxysporum* f *dianthi*. Affected plants wilt rapidly and the leaves become either yellow or gray-green and then straw-colored. In both cases a brown discoloration can be seen in the inner tissues of affected stems. Prevent these diseases by using sterilized pots and compost. Drenching with benomyl may save mildly affected plants. Destroy severely affected plants and sterilize the greenhouse bench or floor on which the plants were standing. Do not propagate from diseased plants. To reduce the spread of wilt drench the remaining plants with a solution of benomyl or thiophanate-methyl, repeating two weeks later.

Stems or crowns with pests visible

Scale insects such as hemispherical scale (*Saissetia coffeae*) and soft scale (*Coccus hesperidum*) encrust the stems of many different plants. The former have red-brown convex shells about $\frac{1}{4}$ in in diameter, while the latter have yellow-brown, flat, oval shells of the same length. The insects live underneath these shells and feed on sap. Once they have found a suitable feeding place they do not move. Control them by spraying plants with malathion or nicotine three times at two week intervals.

Mealybugs (*Pseudococcus* spp) are gray-white soft-bodied insects that grow up to $\frac{1}{4}$ in long. They infest cacti, succulents and many other plants, and secrete white, waxy fibers that cover the mealybug colonies and their egg masses. Control them by spraying with malathion or nicotine. Thorough applications are necessary because mealybugs tend to live on relatively inaccessible parts of the plant, and two or more sprays at two week intervals may be needed. On plants that are liable to be damaged by insecticides, such as *Crassula* and ferns, the old remedy of dabbing mealybugs with a brush dipped in methylated spirit is still effective.

Stems galled

Leafy gall (*Corynebacterium fascians*) affects mainly pelargoniums and chrysanthemums, and shows as a mass of abortive and often fasciated (flattened) shoots at soil level. Destroy affected plants and sterilize pots and the greenhouse bench on which the plants were standing. Do not propagate from diseased plants.

Flower buds dropping

Bud drop affects stephanotis, gardenias, hibiscus and camellias. It is caused by the soil being too dry at the time the buds were beginning to develop. Prevent this trouble by ensuring that the compost never dries out. Gardenias may also lose their buds if the atmosphere is too dry. Prevent this by syringing the plants in the morning and evening during warm sunny weather except when the flowers are open, otherwise they will discolor. Over-watering can also cause bud drop of gardenias.

Flowers discolored

Thrips (various species) are thin, black or yellow insects, about $\frac{1}{10}$ in long, that suck sap from the petals of carnation, chrysanthemum, cyclamen and other plants. The petals develop white flecks where the thrips have fed. Control them by spraying thoroughly with either nicotine or pyrethroids. Care needs to be taken since flowers may be marked with insecticides, so spray when the plants are not exposed to bright sunlight or high temperatures.

Viruses such as cucumber mosaic and tomato spotted wilt can cause spotting or streaking of flowers, which may also be distorted. Most frequently affected are chrysanthemums and bulbous plants, especially lilies and cyclamen. Destroy affected plants.

Flowers spotted or rotting

Gray mold (*Botrytis cinerea*) frequently attacks the flowers of cyclamen and chrysanthemums. For symptoms and treatment, see under Stems or crowns rotting, above.

Pests in or on the soil

Black vine weevil grubs (*Otiorhynchus sulcatus*) are plump white grubs, up to $\frac{1}{2}$ in long, with light brown heads. For symptoms of attack, and treatment, see under Roots and tubers eaten in the Bulbous plants section.

Fungus gnats or sciarids (various species) are small gray-black flies that run over the soil surface of pot plants or fly slowly around them. Their larvae are thin white maggots up to $\frac{1}{4}$ in long with black heads. They live in the soil and feed mainly on rotting plant material but they sometimes damage the roots of seedlings and plants that are in poor health. They may also tunnel into the base of soft cuttings and cause them to rot. Control the adult flies by spraying with a pyrethroid compound. Against the larvae, mix some diazinon granules into the compost as soon as they are noticed.

Springtails (various species) are white, soil-dwelling insects, about $\frac{1}{10}$ in long. They are found especially in peat-based composts, and are distinguished by their habit of jumping when exposed on the soil surface, as occurs after watering. They cause no damage to plants and no controls are required.

Bulbs, corms and rhizomes 1

This section covers the following plants; it excludes those that are only grown under glass.

Acidanthera
Allium
Anemone
Bluebell
Chionodoxa

Crocus
Dahlia
Eranthis (winter aconite)
Freesia
Gladiolus
Hyacinth
Iris
Lily
Muscari

Narcissus
Scilla
Snowdrop
Solomon's seal
Tulip

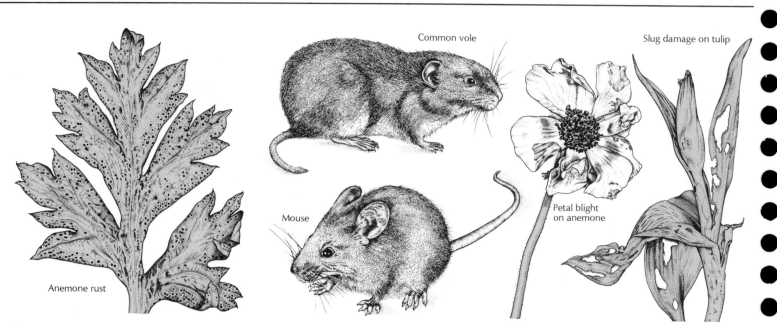

Common vole

Mouse

Anemone rust

Petal blight on anemone

Slug damage on tulip

GENERAL PESTS

The pests in this section may attack any out-door plant grown from a bulb, corm, tuber or rhizome.

Bulbs and corms eaten

Mice, voles and squirrels dig up and eat bulbs and corms between the fall and spring, especially those that have been planted recently. Firmly press down the soil after planting because animals are able to locate bulbs and corms by digging where the soil is soft. Dusting bulbs and corms with alum or a proprietary animal repellant before planting may help to deter rodents. Mouse-traps can be set for mice and voles but take care to cover the traps so that birds and pets cannot set them off. Traps and cages for capturing squirrels are available from gun shops and these require similar care in their use.

Slugs (*Milax* spp) live below the soil surface and eat holes in bulbs and, less frequently, corms. They also eat the emerging leaves and flower shoots. This causes ragged holes to appear in the foliage, and sometimes prevents the flowers from developing. Slugs are difficult to control because they do not often come into contact with slug pellets. The best time to take control measures is during warm weather after heavy rain since this is when slugs come up to the surface. Methiocarb slug pellets will help to reduce the slug population, as will drenching the soil with liquid metaldehyde.

Leaves and flowers with pests visible

Aphids (various species) suck sap from the flowers and leaves, and can also damage plants by transmitting virus diseases. Control them by spraying thoroughly with pirimicarb, dimethoate, formothion or malathion. Aphids may also infest bulbs and corms while they are in storage over winter. Virus transmission can be very rapid under such circumstances and the bulbs and corms may be so weakened by aphids that subsequent growth is poor. Dip infested bulbs and corms in a spray-strength solution of one of the above insecticides for 15 minutes. After dipping, the bulbs or corms should either be planted or, if they are to be returned to store, they should be dried first.

SPECIFIC PESTS AND DISEASES

The pests and diseases in this section are frequently specific to a particular host plant. Therefore, the problems dealt with in this section are grouped according to host rather than by symptom.

Allium

Rust (*Puccinia allii*) affects many species of *Allium* and shows as elongated yellow or red-yellow pustules on the leaves. Remove affected leaves, and in severe cases destroy diseased plants.

White rot (*Sclerotium cepivorum*) rots the roots and the base of the bulb, and covers the affected tissues with a white fluffy growth of fungus. Destroy diseased plants and do not plant alliums on the same site.

Anemone

Downy mildew (*Peronospora ficariae*) shows as a pale fine mold on the lower surfaces of leaves, which tend to roll upwards and may become brown or black. Remove affected leaves and spray with maneb, mancozeb or zineb.

Gray mold (*Botrytis cinerea*) may cause rotting of the buds and flowers in wet weather, especially during the winter. Affected tissues develop a gray velvety mold. Remove diseased parts and spray with captan, thiram, benomyl or thiophanate-methyl. Note that regular use of the last two fungicides may lead to the development of resistant strains of the fungus.

Leaf spot (*Septoria anemones* and *Phyllosticta anemonicola*) causes leaves to develop dark, sharply defined spots, which are dry and sunken. Alternatively, complete browning of the leaves may occur, particularly in winter. Pinpoint-sized black fruiting bodies of the fungus may show on the spots. Remove and burn diseased leaves and spray with captan or zineb.

Rust (*Tranzschelia pruni-spinosae*) shows on leaves and stems as small yellow cups bearing fungus spores. Affected leaves become distorted, are thicker and less divided than normal, and have long thick leaf stalks. Diseased plants do not flower and should be dug up and burned.

Smut (*Urocystis anemones*) affects leaves and stems, causing blister-like swellings that burst open to release a mass of black powdery spores. Destroy diseased plants.

Petal blight (*Itersonilia perplexans*) causes small oval translucent spots on the outer florets and, in wet weather, spreads rapidly.

Where this disease has been troublesome spray the flower buds with zineb. Cut off and burn infected blooms.

Viruses such as arabis mosaic and cucumber mosaic cause plants to become stunted and bear yellow or brown distorted leaves. The flowers are small and have a poor color. Destroy affected plants, and do not replant anemones on the same site.

Bluebell, chionodoxa, muscari and scilla

Rust (*Uromyces muscari*) causes yellow spots to develop on both sides of the leaves of bluebell, muscari and scilla. These spots later burst open and produce powdery masses of dark brown fungus spores. Spray affected plants with zineb.

Smut (*Ustillago vaillantii*) affects chionodoxa, muscari and scilla, converting the anthers, and less frequenty the ovaries, in the flowers to masses of black powdery spores. Destroy affected plants.

Crocus

House sparrows (*Passer domesticus*) attack the flowers, sometimes tearing them to shreds. The damage is often localized, with crocuses in nearby gardens remaining un-scathed. Also, some colors may be attacked

Bulbs, corms and rhizomes 2

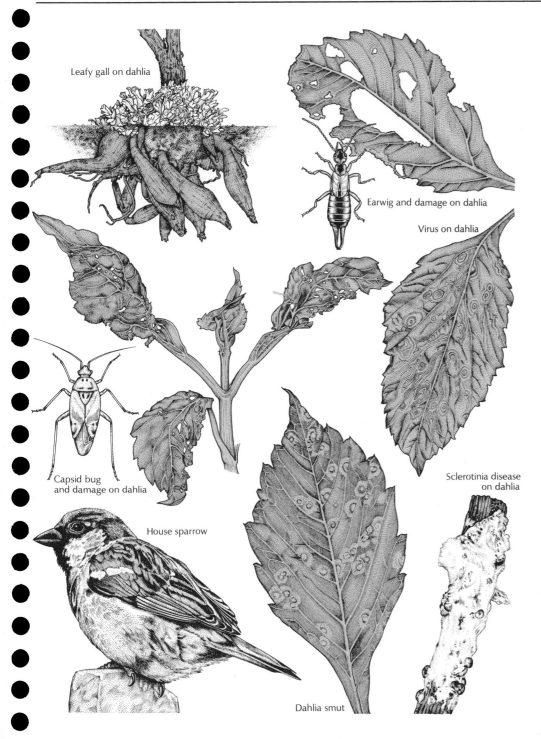

Leafy gall on dahlia

Earwig and damage on dahlia

Virus on dahlia

Capsid bug and damage on dahlia

House sparrow

Sclerotinia disease on dahlia

Dahlia smut

in preference to others, but no flower color is immune and the most susceptible color may vary from year to year or between different locations. Sparrows rip the flowers but do not eat them, therefore bird repellant sprays are unlikely to be effective because the birds do not have much opportunity to taste the repellant substances. Netting ensures that birds are kept away from the flowers but this can be unsightly. An alternative is to place sticks among the plants and criss-cross black cotton thread between the sticks just above the flowers. Sparrows have difficulty in seeing the threads and are frightened off when they fly into them.

Crinkling of leaves is caused by a check in growth, usually from the soil being too dry. The flower buds of affected plants may turn brown and fail to open. Prevent these troubles by watering before the soil dries out completely. Occasionally however, leaf crinkling occurs as a result of frost damage and nothing can be done to prevent it. For other diseases of crocus see Narcissus basal rot and the section on gladiolus.

Dahlia

Blackfly (*Aphis fabae*) and other aphids form dense colonies on the shoots and flower stems between late May and August. This can lead to reduced vigor and smaller flowers. Dahlias suffer from a number of aphid-spread virus diseases, so it is worth while keeping aphids under control by spraying with dimethoate, formothion, pirimicarb or malathion as soon as they are noticed.

Capsid bugs (*Lygus rugulipennis* and *Lygocoris pabulinus*) are brown or green insects about $\frac{3}{8}$ in long that suck sap from young leaves, shoot-tips and flower buds. Their damage to young leaves only shows some weeks later when the leaves have expanded and developed a tattered, distorted appearance due to the presence of innumerable small holes. Where the bugs have been feeding on the flower buds, the flowers open unevenly and become distorted. Capsid bugs are active throughout the summer but most damage occurs in late May to July. As soon as the bugs, or signs of their damage, are seen spray the plants with dimethoate, formothion, fenitrothion or malathion.

Earwigs (*Forficula auricularia*) feed nocturnally and so may go unnoticed unless the plants are examined by torchlight. During the day earwigs hide between the petals or the folds of opening leaves. Their feeding causes irregular holes to appear in the foliage, and the flowers become spoiled with the earwigs' droppings. They also chew the petals, causing them to turn brown. Trap the earwigs by placing rolls of corrugated cardboard, or clay pots loosely stuffed with straw, among the plants. The earwigs will use these places as shelter during the day, when they can be removed and destroyed. During the flowering period heavy infestations will require the use of insecticides, in which case apply carbaryl dusts at dusk on warm evenings. The soil surface below the plants should also be treated.

Gray mold (*Botrytis cinerea*) can cause flowers to rot in wet summers. Affected tissues develop brown patches which later become covered with a gray velvety fungal growth. Remove diseased flowers; prevent attacks of gray mold by spraying the flower buds with zineb or benomyl.

Petal blight (*Itersonilia perplexans*) may also affect the flowers in a wet season. Small brown water-soaked spots develop on the outer florets, then spread inwards until the whole bloom becomes brown and rotten. The disease is usually followed by gray mold. Where this disease is known to occur spray the flower buds with zineb or benomyl.

Smut (*Entyloma dahliae*) affects the lower leaves of bedding dahlias, causing circular brown spots to appear. As it spreads upwards the spots coalesce and the leaves wither. Cut off and burn affected leaves and spray with a copper fungicide, repeating the treatment once or twice at two week intervals if necessary.

Leafy gall (*Corynebacterium fascians*) affects dahlias at ground level, producing a mass of often flattened shoots that fail to develop. Destroy affected plants.

Viruses such as cucumber mosaic, dahlia mosaic and tomato spotted wilt commonly infect dahlias, causing stunting of plants whose leaves show yellow or brown rings or spots and may also be distorted. Destroy affected plants.

Bulbs, corms and rhizomes 3

Sclerotinia disease (*Sclerotinia sclerotiorum*) causes the stems to decay and the parts above to collapse. A fluffy white fungal growth develops within the stems, and the tubers may be affected in store. Hard black resting bodies of the fungus, $\frac{1}{2}$ in or more in length, develop on the diseased tissues and fall into the soil or to the floor. They remain dormant there until the following season when they germinate and produce spores that cause new infections. Destroy any plant or tuber showing these symptoms, preferably before the resting bodies are formed. Grow dahlias on a fresh site the following season.

Eranthis (winter aconite)
Smut (*Urocystis eranthidis*) affects the leaves and leaf stalks, causing blister-like swellings that burst open to release a mass of spores resembling soot. Destroy affected plants.

Gladiolus, freesia, crocus and acidanthera
Gladiolus thrips (*Taeniothrips simplex*) are thin, yellow or dark brown insects, up to $\frac{1}{10}$ in long. They feed by sucking sap from the leaves and flowers of gladiolus and are usually more troublesome in hot dry summers. Their feeding causes a silvery mottled discoloration on the foliage. The flowers may be similarly marked, but if the flower buds are heavily infested the petals turn brown and the buds fail to open. Check gladioli at intervals during the growing season and if thrips damage is noticed spray plants with dimethoate, formothion or carbaryl. Avoid treatment once the flowers have started to open because the petals can be marked by insecticides. Thrips can over-winter on the corms, therefore dust them with carbaryl before putting them into storage.
Core rot (*Botrytis gladiolorum*) is common on gladiolus but less frequent on acidantheras, crocus and freesias. It is usually a storage disease that starts at the central core of the corm and spreads outwards. The corm becomes dark and spongy, and eventually rots completely. If affected corms are planted they either fail to grow or produce weak shoots that turn yellow and die. Less badly diseased corms produce leaves showing small brown spots with red margins. In wet weather, the leaves may decay and become covered with a gray mold. The flowers may also become spotted. Prevent storage rot by drying the corms thoroughly and dusting with quintozene.

Dry rot (*Stromatinia gladioli*) is usually introduced into a garden on slightly affected corms. The foliage of these corms turns yellow prematurely and dies, and the leaf sheaths decay at ground level, sometimes causing plants to collapse. Therefore, remove and burn affected plants, and at lifting time clean the corms roughly and destroy any that show black patches. Dust with quintozene before storing, which should be in a cool but frost-free place with a dry atmosphere. Prior to planting, or immediately after lifting, dip the corms for 15 minutes in a solution of benomyl, or place them in a solution of captan for one hour. Rake quintozene or DCNA into the soil before replanting.

Hard rot (*Septoria gladioli*) causes minute brown spots on upper and lower leaf surfaces. Small black fruiting bodies of the fungus develop on these spots and produce spores that spread the disease rapidly in wet weather. Destroy diseased plants and prevent storage rot by dusting or dipping the corms as recommended for dry rot (see above).

Scab (*Pseudomonas marginata*) shows on the leaves as red-brown specks which later enlarge and darken. In wet weather decay of the tissues at ground level may occur, causing the top growth to fall over. Towards the base of the corm sunken round craters develop, each of which has a prominent raised rim and, often, a shiny coating. Burn diseased plants and corms. Do not grow gladioli on the same site again because the bacteria can survive from one season to another in the soil. The treatment recommended above for dry rot gives some control over scab.

Yellows (*Fusarium oxysporum* f *gladioli*) shows first as yellow stripes between the veins. The leaves then become completely yellow and die back. If the corm of an affected plant is cut across, the base can be seen to be dead with brown vascular strands radiating from it. As the corm symptoms appear only within the tissues the disease is difficult to detect, and contamination of the soil occurs if even slightly infected corms are planted. The only

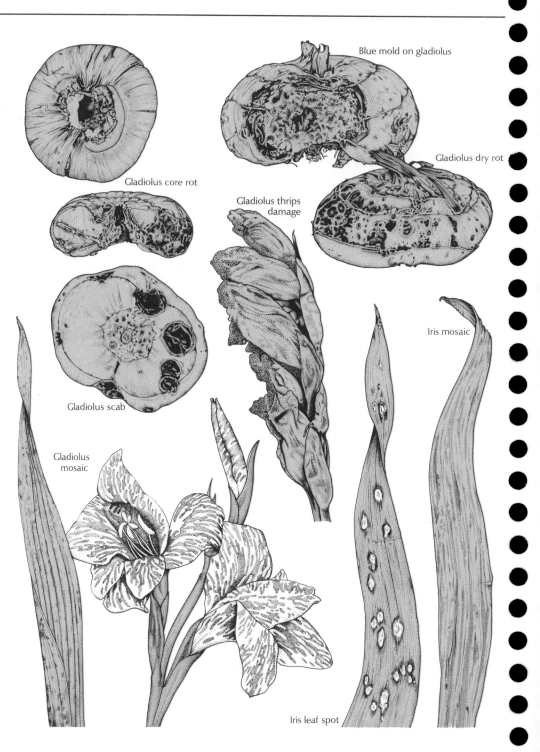

Blue mold on gladiolus

Gladiolus dry rot

Gladiolus core rot

Gladiolus thrips damage

Iris mosaic

Gladiolus scab

Gladiolus mosaic

Iris leaf spot

Bulbs, corms and rhizomes 4

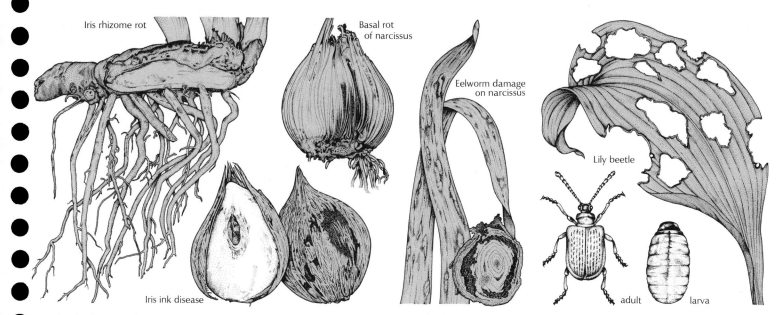

Iris rhizome rot

Basal rot of narcissus

Eelworm damage on narcissus

Lily beetle

adult   larva

Iris ink disease

method of control is to remove and burn diseased plants as soon as the first symptoms are seen. Grow gladiolus corms on a fresh site each year.

Blue mold (*Penicillium* sp) is a storage disease that causes the corm to develop a few sunken lesions, $\frac{1}{2}$ in or more across. These are red-brown and bear buff or pink resting bodies of the fungus. In moist conditions a blue mold develops on the lesions, and the corm decays. Prevent blue mold by drying off corms well. Store only healthy undamaged corms in a cool dry place, and dust or dip as recommended for dry rot. Examine corms in store at regular intervals and discard any showing signs of rotting.

Viruses such as cucumber mosaic and bean yellow mosaic cause white blotches on the flowers, which may also be distorted, and yellow streaks or spots on the leaves. Prominent yellow rings or ring and line patterns may also appear on the leaves. Freesias usually show red-brown streaks on the leaves. Destroy plants showing any of these symptoms.

Hyacinth

Gray bulb rot (*Rhizoctonia tuliparum*, syn *Sclerotium tuliparum*) is a soil-borne disease.

The fungus attacks the bulb at the nose, causing a dry gray rot which either prevents the shoot from growing much above soil level or kills the bulb before any shoots develop. Dense fungal growths develop on the rotting tissues of the bulb and quickly produce large black resting bodies that contaminate the soil. These germinate the following season and infect new bulbs, resulting in bare patches in the flower bed. Remove and burn diseased plants together with any debris of diseased bulbs found in the gaps in the bed, and also some of the soil surrounding them. Plant only healthy bulbs after dusting them with quintozene and raking this fungicide into the soil. Susceptible plants include colchicum, crocus, narcissus, gladiolus, ixia, scilla, fritillary and tulip, as well as hyacinths. Grow all these on a fresh site each season to prevent the disease from building up.

Iris (rhizomatous)

Leaf spot (*Mycosphaerella macrospora*) shows on the leaves as brown oval-shaped spots that coalesce and sometimes kill the leaf. It is most troublesome in wet weather, particularly in the spring and fall. Cut off and burn affected leaves and spray with maneb or zineb plus a spreader-sticker.

Rhizome rot (*Erwinia carotovora*) causes the rhizomes to develop a yellow slimy foul-smelling rot. The leaves turn brown and die back from the tips and the fan collapses at ground level. Heavy losses may occur in a wet season. Discourage the disease by growing iris in well drained, properly cultivated soil. Prevent infection by planting shallowly and carefully, and by controlling slugs. Destroy affected plants. Cut out affected parts of less severely diseased plants as soon as they are seen, and dip the knife frequently in a disinfectant during the operation. Dust the wounds, rhizomes and soil with bordeaux powder to prevent reinfection.

Iris scorch is a disorder that is possibly caused by either unsuitable soil conditions during the winter or drought in the spring. The leaves become red-brown and, as they die, they wilt and bend over. Eventually all the leaves in a clump may be affected. The rhizomes remain firm and healthy but the roots of a diseased plant are nearly all dead and many are reduced to hollow tubes of outer skin with a stringy core. Affected plants may recover if treated early. Lift them and remove all dead leaves and rotten roots. Wash the rhizomes well, adding some potassium permanganate crystals to the water until it is

pale pink. Replant the rhizomes in well drained soil and water with Cheshunt compound.

Viruses such as arabis mosaic and cucumber mosaic cause yellow streaks and spots on the leaves. Severely affected plants may be stunted and should be destroyed. Wash hands and tools after handling the plants.

Iris (bulbous)

Ink disease (*Drechslera iridis*) causes black crusty patches to appear on the outer scales of the bulb. The inner tissues then decay to leave an empty shell containing a small quantity of black powder. Fungus spores develop on the blackened scales and, if slightly affected bulbs are planted, the fungus can spread from diseased to healthy bulbs. In a wet season black spots develop on the leaves, sometimes affecting the whole leaf surface, and also occasionally on the flowers. Remove and burn the blackened covering scales and discard all bulbs in which the inner scales are also affected. In areas where this disease is known to be troublesome spray developing leaves with mancozeb. At the end of the season, cover the plants with straw and burn it off on a windless day.

Leaf spot (*Mycosphaerella macrospora*) can be troublesome in two-year-old plants. For symptoms and treatment see Leaf spot on rhizomatous irises.

Iris mosaic virus causes yellow streaks and spots on the younger leaves. Affected plants may be stunted and usually produce broken flowers showing dark streaks. The variety 'Wedgwood' is particularly susceptible to this disease. Destroy affected plants. Wash hands and tools after handling the plants.

Lily

Lily beetles (*Lilioceris lilii*) can cause great damage. Both the adult beetles and their larvae feed on the foliage, flowers and seed capsules, and may completely defoliate plants. The beetles are $\frac{3}{8}$ in long and bright red with black legs, heads and underparts. Their larvae are plump and orange-red, but are usually disguised by a covering of black, slimy excreta. Attacks begin in late April or May when the adult beetles emerge from the soil. Spray or dust lilies with malathion,

Bulbs, corms and rhizomes 5

or pirimiphos-methyl as soon as the pests are seen.

Lily disease or leaf blight (*Botrytis elliptica*) causes oval-shaped water-soaked spots on the leaves. In humid weather these spread rapidly, become brown and the leaves may rot. The fungus sometimes spreads from the leaves and causes flowers to rot or stems to topple. *L. candidum* and *L. × testaceum* suffer most from this disease but it is also fairly common on *L. regale*, though the symptoms are usually less severe. Seedlings of all types of lily can also be attacked. Cut off and burn diseased top growth to eliminate the over-wintering stage of the fungus by which reinfection may occur. Infection can also be caused by spores produced on blighted rosette leaves of *L. candidum*, so check over-wintering leaves occasionally and remove any that show spots. Spray with benomyl or a copper fungicide soon after new leaves appear in the spring, repeating at 10–14 day intervals until flowering. In a wet season it may be necessary to start spraying again when flowering has finished. Grow lilies on a good, well drained site where there is free circulation of air so that the leaves can dry off quickly after rain or dew.

Basal rot (various fungi) causes the roots and tissue at the base of the bulb to rot, resulting in discoloration of foliage and ceasing of the top growth. Destroy badly affected bulbs. In less severe cases cut out diseased roots, basal tissues and scales. Dust the remainder of the bulbs with quintozene and rake this fungicide into the soil before replanting on a fresh site. Should further trouble occur dip surviving bulbs for 30 minutes in a 2 per cent solution of formalin, or for 15 to 30 minutes in a solution of benomyl made at the rate of $\frac{1}{2}$ oz in 2 gal of water. Do not use benomyl every year since it could lead to the development of blue mold (see page 67).

Viruses such as cucumber mosaic, lily mottle and lily rosette cause irregular chlorotic streaks, slight mottling or rosetting of the leaves, spotting and distortion of the flowers and stunting of plants. The last-named virus also has a marked effect on the bulbs, which are smaller and flatter than usual and have a tendency to split up. Destroy affected plants and wash the hands with soapy water after

Solomon's seal
sawfly caterpillars

handling a diseased plant. Do not grow Rembrandt tulips close to lilies because these contain lily mottle virus.

Narcissus

Narcissus eelworms (*Ditylenchus dipsaci*) are microscopic, worm-like animals that live inside the bulb, foliage and flower stems. Infested plants produce short distorted leaves which may have rough yellow patches on them. Flower stems are similarly distorted or fail to emerge. Eelworms spread through the soil or by leaf contact so that each year the number of damaged bulbs gradually becomes larger. Badly infested bulbs are killed. If a bulb is cut in half transversely, infested tissues can be seen as brown concentric rings. None of the chemicals available to amateur gardeners controls eelworms and the only measure is to dig up and burn infested bulbs. Other narcissi growing within a yard of such plants should also be removed since they may be infected. Host plants of this pest include snowdrop, bluebell, hyacinth, onion, beans, peas, strawberries and some common weeds. Do not grow these in infected soil for at least two years.

Large narcissus bulb flies (*Merodon equestris*) lay their eggs during late spring and early

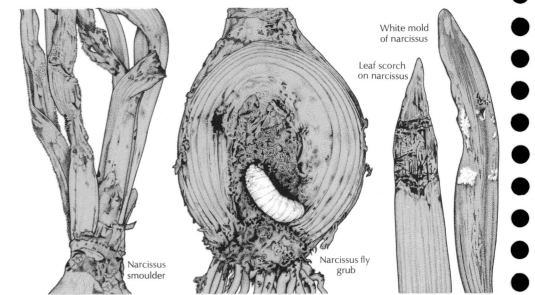

White mold
of narcissus

Leaf scorch
on narcissus

Narcissus
smoulder

Narcissus fly
grub

summer near the neck of narcissus bulbs. The maggots tunnel into the bulb where they destroy the central portion before they are fully fed in the following spring. Often the bulb is killed but it may manage to produce a few thin leaves without any flowers. There is usually only one maggot in a bulb. When fully grown it is stout, $\frac{3}{4}$ in long and a dull cream color. The center of the bulb becomes filled with the grub's soft muddy excreta. Deter the female flies from laying eggs by raking soil up around the bulbs as the foliage dies down. These flies are difficult to control but help to reduce infestation by dusting the dying foliage and the soil around bulbs with malathion at 10–14 day intervals between late April and June. Snowdrop, bluebell and other bulbs of the Amaryllidaceae family may also be attacked.

Basal rot (*Fusarium oxysporum* f *narcissi*) usually occurs in warm weather when the roots are dying back naturally. If an infected bulb is not lifted at the end of the season it will rot in the soil and the disease will spread gradually to adjacent bulbs. Infected bulbs that are lifted and stored show signs of rotting after a month. The basal plate develops a brown discoloration and the bulb feels soft in this area. Eventually it shrivels up and

becomes hard. Destroy all affected bulbs, including those that feel soft. If basal rot has caused losses in previous seasons lift bulbs in June before the soil becomes hot. Within 48 hours of lifting dip them for up to 30 minutes in a solution of benomyl made at the rate of $1\frac{1}{2}$ oz in 6 gal of water, keeping the suspension agitated. Dry off the bulbs before storing in an airy, cool, but frost-proof place. Plant only firm healthy bulbs in early fall, in well drained soil, and change the site for daffodils and other narcissi every few years.

Smoulder (*Sclerotinia narcissicola*) causes stored bulbs to decay and foliage and flowers to rot in cold wet seasons. Affected parts become covered with a mass of gray airborne spores and the leaves are killed rapidly. In warmer, drier conditions infection is usually restricted to one or two leaves and the plants flower normally. Spray as soon as the disease is seen using benomyl or zineb and repeat once or twice before flowering. Dipping with benomyl as for basal rot (see above) will also control this disease. Destroy any bulbs that bear the small black flattened resting bodies of the fungus.

Leaf scorch or tip burn (*Stagonospora curtisii*) attacks the developing leaves, giving them a scorched or burnt appearance at the

Bulbs, corms and rhizomes 6

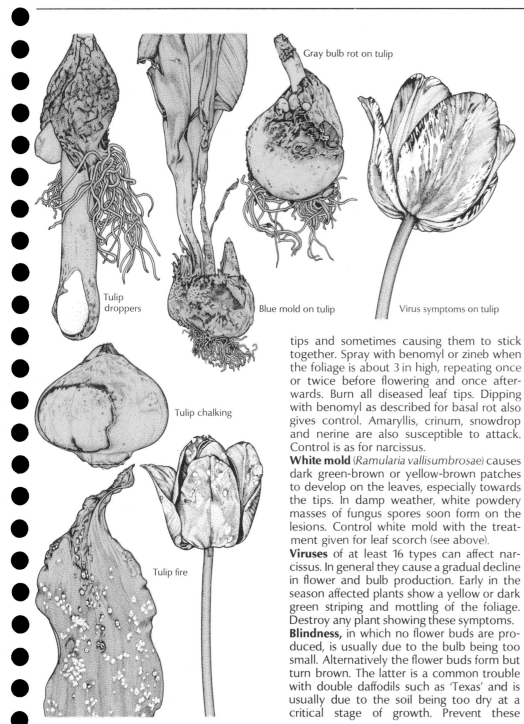

Gray bulb rot on tulip

Tulip droppers

Blue mold on tulip

Virus symptoms on tulip

Tulip chalking

Tulip fire

tips and sometimes causing them to stick together. Spray with benomyl or zineb when the foliage is about 3 in high, repeating once or twice before flowering and once afterwards. Burn all diseased leaf tips. Dipping with benomyl as described for basal rot also gives control. Amaryllis, crinum, snowdrop and nerine are also susceptible to attack. Control is as for narcissus.

White mold (*Ramularia vallisumbrosae*) causes dark green-brown or yellow-brown patches to develop on the leaves, especially towards the tips. In damp weather, white powdery masses of fungus spores soon form on the lesions. Control white mold with the treatment given for leaf scorch (see above).

Viruses of at least 16 types can affect narcissus. In general they cause a gradual decline in flower and bulb production. Early in the season affected plants show a yellow or dark green striping and mottling of the foliage. Destroy any plant showing these symptoms.

Blindness, in which no flower buds are produced, is usually due to the bulb being too small. Alternatively the flower buds form but turn brown. The latter is a common trouble with double daffodils such as 'Texas' and is usually due to the soil being too dry at a critical stage of growth. Prevent these

troubles by lifting and dividing bulbs annually, by feeding naturalized bulbs, and by watering double varieties before the soil dries out completely. Spraying with a foliar feed after flowering will also help to prevent blindness.

Snowdrop
Gray mold (*Botrytis galanthina*) causes the leaves and flower stalks to rot and become covered with a gray velvety fungal growth. Small black resting bodies of the fungus develop on the bulb, which then decays. Destroy infected clumps and treat the remainder with quintozene, captan, benomyl, thiophanate-methyl or thiram.

Solomon's seal
Solomon's seal sawfly (*Phymatocera aterrima*) are black flies about ½ in long with two pairs of dark wings. They emerge when the plants are in flower and lay a batch of up to 20 eggs in the stems. These hatch into gray-white caterpillars with black heads. By the time these larvae are fully grown, when they are 1 in long, they may have completely defoliated the plant. They then go into the soil where they over-winter before pupating in the late spring. Prevent damage by spraying the plants with derris or pirimiphos-methyl when young larvae are first noticed, which is in late May to June.

Tulip
Tulip fire (*Botrytis tulipae*) is a common and serious disease of tulips. A brown scorch shows at the tip of the young leaves and severely affected plants may be crippled. In moist weather a gray mold develops and causes rotting of the affected tissues. The disease then spreads to produce small brown spots on leaves and flowers, which may also rot. Small black resting bodies of the fungus are produced on the bulbs. These bulbs may also rot, and should be destroyed. Dig up and burn crippled or rotting plants. To control the disease rake quintozene into the soil and dust the bulbs with it before planting. When the leaves are 2 in high, spray with dichlofluanid, mancozeb, maneb, benomyl or zineb and repeat at 10–14 day intervals until flowering. Within 24 hours after lifting, dip the bulbs for 15 minutes in a solution of

benomyl made at the rate of 1 oz in 4 gal of water, keeping the suspension agitated. Dry the bulbs before storing.

Gray bulb rot (*Rhizoctonia tuliparum*) is a soil-borne disease. For symptoms and treatment, see under Hyacinth.

Blue mold (*Penicillium* sp) may develop on injured bulbs, particularly in storage. If slightly infected bulbs are planted they may produce good plants, but in wet weather the fungus may cause rotting of the bulbs resulting in discoloration or death of the leaves, or poor growth. A blue-green mold develops on the affected tissues. Do not plant severely affected bulbs. In less serious cases cut out the affected tissues and dust the bulbs with quintozene before planting. Control pests such as slugs that injure bulbs and make them susceptible to infection.

"Droppers" are swollen contractile roots that show as long, elliptical structures growing from the top of the bulb down into the soil. They will eventually form a bulb but it will take several years to reach flowering size, so affected bulbs are not usually worth keeping. The phenomenon is caused by planting bulbs too shallowly, or by the soil being too dry. Ensure that bulbs are planted at the correct depth, and water in dry periods.

Blindness is usually caused by the bulb being too small. If the flower buds are formed but turn brown it is usually due to the soil being too dry at a critical stage of growth, or by storing the bulbs in too warm a place before planting. Prevent these troubles by planting only large bulbs or by feeding annually, and make sure that the soil never dries out.

Chalkiness and hardness of bulbs is caused by lifting too early and incorrect storing. Prevent this by lifting bulbs after the foliage has died down, or if they must be lifted with green leaves, heel them in somewhere in the garden until the leaves have died. Store them in a cool airy place. Destroy infected bulbs since there is no cure.

Viruses cause the flower color to break with the petals having either white streaks or darker streaks than the normal color. Other symptoms produced are irregular brown or white streaks, or brown spots, on the leaves. Destroy any plant showing these symptoms unless Rembrandt tulips are desired.

Plants raised from seed 1

This section covers the following plants; it excludes those that are only grown under glass.
Ageratum
Althaea (hollyhock)
Antirrhinum
Aquilegia (columbine)
Aubrieta
Calceolaria
Calendula (marigold)
Callistephus (China aster)
Campanula (Canterbury bell)
Celosia
Centaurea (cornflower)
Cheiranthus (wallflower)
Clarkia
Dianthus (sweet william)
Dimorphotheca
Godetia
Helianthus (sunflower)
Ipomoea
Lathyrus (sweet pea)
Lavatera
Limonium (statice)
Lobelia
Lunaria (honesty)

SEEDLINGS
This section covers the period of plant growth between germination and the emergence of true leaves.

Seedlings eaten
Slugs, woodlice and millipedes can destroy plants by eating the foliage before the seedlings have a chance to become established. Slugs are the most destructive; woodlice and millipedes only become troublesome when they are present in large numbers. Slug pellets containing methiocarb give some additional protection against woodlice and millipedes, but if these pests are very numerous methiocarb will not control them and they must be tolerated.

Seedlings collapsing
Damping off is usually due to species of the soil- and water-borne fungi *Phytophthora* and *Pythium*. Seedlings of antirrhinum, sweet peas, lobelia, stock and zinnia are particularly susceptible to infection, and collapse at ground level. Prevent infection by sowing thinly, since the disease is encouraged by overcrowding, and by using sterilized soil or compost of a good tilth. Over-watering can also induce damping off, so water carefully with clean water. Give adequate illumination but not too much heat. Check slight attacks by watering with ethazol, Banrot or Cheshunt compound after all dead seedlings have been removed and destroyed.

Seedlings rotting
Seedling blight (*Alternaria zinniae*) affects zinnias, causing red-brown spots with gray centers to develop on the leaves and dark brown canker-like areas on the stems. Affected seedlings collapse and die, and should be destroyed. Prevent infection by dusting seed with a captan or thiram seed dressing and spraying seedlings with zineb or mancozeb.
Leaf and stem rot (*Alternaria tenuis*) affects lobelias. Pale spots develop on seedling leaves to give a scorched effect, and in severe cases the seedlings damp off. Destroy affected plants. Prevent infection by dusting seed with a captan or thiram seed dressing and by spraying with thiram, zineb or mancozeb.

Leaves discolored
Downy mildew is caused by *Peronospora antirrhini* on antirrhinums, and *Peronospora parasitica* on stocks and wallflowers. White mealy or downy tufts of fungal growth develop on the leaves, which become blotched on the upper surface. In severe cases the leaves and stems may be distorted and also moldy, and the plants receive a severe check in growth. Prevent these diseases by sowing seed thinly in sterilized, well drained soil or compost, and ventilate carefully to reduce the humidity. Remove all diseased leaves and spray plants with mancozeb, chlorothalonil or zineb.

MATURE PLANTS
This section covers the period from the emergence of true leaves onwards.

Leaves with pests visible
Aphids (many species) suck sap from the leaves and stems of most plants. Besides spreading virus diseases they cause various symptoms such as stunted growth, poor flowering and curled leaves. They also soil the foliage with a sticky substance known as honeydew. This allows the growth of a black sooty mold on the leaves. Control aphids by spraying the plants thoroughly with any of the insecticides pirimicarb, dimethoate, formothion or pirimiphos-methyl.

Leaves eaten
Slugs, snails and caterpillars make irregular holes in the leaves. Slug and snail damage can be recognized by the silvery slime trail that they leave on the foliage. Scatter slug pellets among the plants during periods of warm, damp weather since this is when the greatest number of slugs and snails will be feeding. If caterpillars are causing the damage they can usually be found on the plants, although it may be necessary to examine them by torchlight since many are nocturnal in their feeding habits. If the plants are too heavily infested for hand-picking, spray them with fenitrothion or derris.
Earwigs (*Forficula auricularia*) eat the soft parts of leaves until only the network of veins remains. In some years earwigs become very abundant and may attack a wide range of

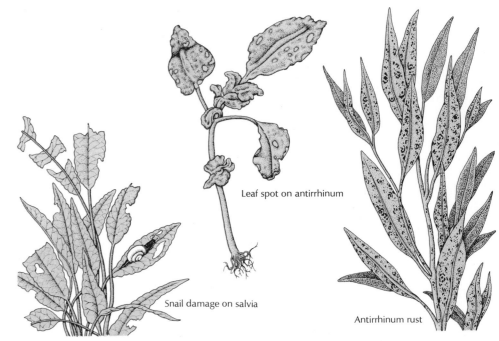

Leaf spot on antirrhinum

Snail damage on salvia

Antirrhinum rust

plants. Like many caterpillars, they feed at night, so a torchlight inspection is necessary to discover the extent of an infestation. If the infestation is heavy spray the plants with fenitrothion or trichlorphon at dusk on warm, still evenings.

Leaves discolored
Cold night temperatures can cause the leaves of ipomoea and sweet peas to become white, those of antirrhinums to become silver, and those of other bedding plants to turn pale yellow. Give some form of protection if frost is forecast. Applications of a foliar feed may help to restore the green color to the foliage.

Leaves spotted
Leaf spot may be caused by a variety of fungi. Small circular spots, usually brown in color, develop on the leaves and may coalesce. Remove and burn affected leaves, and spray with zineb, maneb or mancozeb, repeating once or twice at two week intervals. If the trouble persists the plants may be lacking in vigor due to some trouble at the

roots. In this case, spray plants with a foliar feed, but do not feed with nitrogenous fertilizers.

Leaves blotched
Sweet pea scorch is a physiological disorder, the exact cause of which is not known. Pale brown blotches develop on the leaves and may coalesce. This disorder occurs most frequently on sweet peas grown cordon fashion and spreads upwards, sometimes killing all the leaves. Where this trouble occurs annually, grow sweet peas by the bush method. At the first signs of trouble remove severely affected leaves and spray plants with a foliar feed.
Viruses of many types may affect plants, particularly sweet peas. They cause a yellow mottling or blotching of leaves, which may be distorted if the plants are stunted. Destroy affected plants and control aphids since they spread some of these viruses.

Leaves with visible fungal growth
Rust may be caused by a variety of fungi. On antirrhinums and cornflowers dark brown

Plants raised from seed 2

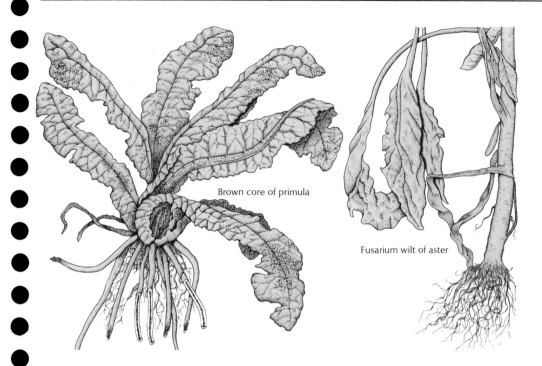

Brown core of primula

Fusarium wilt of aster

masses of spores develop on the leaves and stems, and severely affected plants may be killed. On sweet williams the spore pustules may also be brown, and appear as irregular rings on the leaves. On hollyhocks and lavatera yellow-orange spore-producing pustules develop on the leaves, leaf stalks and stems. Primrose leaves are similarly affected, but only on the undersides. In slight attacks remove affected leaves and spray with zineb, maneb or mancozeb. Destroy severely affected plants. Try to prevent infection by growing rust-resistant varieties of antirrhinums (though the resistance has broken down in many localities), by growing sweet williams fairly hard (that is, do not apply nitrogenous fertilizers or manure) and by raising new hollyhocks from seed every other year.

Powdery mildew of various species can affect a wide range of plants. A white powdery coating develops on the leaves and sometimes the stems. Plants are more susceptible if the soil is too dry, therefore mulch to conserve moisture, and water in dry periods. At the first signs of trouble spray

with benomyl, dinocap, thiophanate-methyl or quintozene, repeating as necessary. Remove severely affected plants and destroy all diseased annuals at the end of the season.

White blister (*Albugo candida*) shows as white glistening pustules on the leaves and stems of aubrieta and honesty. The spore masses may be either single or in concentric rings. Remove and burn affected parts.

Stems or crowns wilting

Pansy sickness is caused by certain soil-borne fungi. The stems rot at ground level, and the top growth wilts, turns yellow and can be lifted easily. The roots may also be killed. Dig up and burn affected plants together with as many roots as possible, and grow pansies and violas on a different site each year. Prevent infection by watering the plants both in the seed boxes and the planting holes with Cheshunt compound, repeating at weekly intervals.

Fusarium wilt of china aster is caused by *Fusarium oxysporum* f *callistephi*. Affected plants wilt, usually just as they are about to flower. The stems blacken just above ground

level or halfway up, and develop a pink fungal growth. Destroy affected plants and grow asters on a new site each year. Where this disease is a recurrent problem, improve the drainage if it is poor, and grow only varieties listed as wilt-resistant in seed catalogs.

Brown core (*Phytophthora primulae*) of primulas causes the roots to rot back from the tips until only a small cluster of very short roots is left. Diseased plants wilt and can be lifted easily. If the remaining roots are split lengthways a brown core will be seen in them. Burn affected plants. Do not grow primulas on the infected site for many years unless the soil has first been sterilized with a 2 per cent solution of formalin, applied at 6 gal per square yard, or dazomet. Soil sterilization is only feasible where primulas are grown in large drifts because these chemicals cannot be used in beds where there are permanent plants.

Crown, foot and root rot may be caused by numerous fungi. The roots, crowns and stem bases of infected plants die, and tissues become brown or black, and rotten. The leaves become discolored and the top growth wilts, often just as the plant is about to come into flower. Plants that are over- or under-watered, or have been poorly planted, are susceptible to infection. Therefore, when planting out, do not insert too deeply and spread the roots out well. Water carefully making sure that the soil is never allowed to dry out, and rotate bedding plants to reduce the possibility of infection. It may be possible to save slightly infected plants by watering with captan, Cheshunt compound or zineb, and by spraying the leaves with a foliar feed. Burn severely diseased plants, and the following season grow a less susceptible type of bedding plant, such as zonal pelargoniums or bedding begonias, in the infected area. In slightly contaminated soils apply a seed dressing of captan. If the infected bed is vacant during the winter, sterilize with a 2 per cent solution of formalin or dazomet as recommended for brown core (see above).

Bean yellow mosaic and common pea mosaic viruses frequently cause sweet peas to wilt. Affected plants first show mottling of the leaves and vertical streaking in the wings of the upper and middle part of the

stem, which is at first light green but later becomes brown. The streaks spread up and down the stem, along the upper side of the leaf stalks and into the leaf veins before the plants collapse. Destroy affected plants and wash hands and tools with soapy water immediately afterwards. Spray regularly with an insecticide to control aphids, since they spread these viruses.

Stems rotting

Leaf spot and stem rot may affect antirrhinums, godetias, stocks and wallflowers. Dark brown spots develop on the stems, eventually girdling and killing them. Small black fruiting bodies of the fungus can be seen on the affected areas on antirrhinums, stocks and wallflowers. These produce spores that spread the disease to the stems of adjacent plants. In the case of antirrhinums the leaves, flowers and fruits may also be infected and develop pale brown circular spots. Remove and burn affected plants before the disease spreads too far, and spray the rest with captan, thiram, zineb, maneb or mancozeb. Control these diseases by dusting seeds with a captan or thiram seed dressing and spraying the seedlings with a solution of thiram, maneb or mancozeb.

Gray mold (*Botrytis cinerea*) may cause rotting of plants in a wet season, affected tissues becoming covered with a gray velvety fungal growth. Cut off and burn affected parts and spray with benomyl or thiophanate-methyl—note that too regular use of these fungicides could lead to a build-up of resistant strains of the fungus. In wallflowers gray mold causes the trouble known as winter killing. Sometimes only the side-shoots are affected but in other cases the main stem may also be injured and the whole plant dies back. Wallflowers are more susceptible to infection if they have grown too soft from excess nitrogen in the soil, or have been planted late or have poor root systems. Apply a dressing of a complete fertilizer to the seedbed, or seed boxes, and also to the soil before planting out. This should be done early so that plants can become fully established before winter.

Sclerotinia rot (*Sclerotinia sclerotiorum*) causes rotting of sunflower stems, but they

Plants raised from seed 3

Virus symptoms on sweet pea

Cutworm damage

Petal blight on cornflower

Pansy rust

do not always show the typical white fluffy growth of the fungus on the outer tissues. More often the fungus develops within the stems where it produces numerous irregular-shaped resting bodies $\frac{1}{2}$ in or more in length. Keep a careful watch on sunflowers during the growing season, particularly if this disease has occurred in previous years or on other crops. Remove and burn any plant showing discolored patches or rotting stems before the resting bodies fall and contaminate the soil. Grow sunflowers on a fresh site the following year.

Stems withering
Cutworms (*Agrotis*, *Euxoa* and *Noctua* spp) are soil-dwelling caterpillars that grow up to $1\frac{1}{2}$ in long and are usually creamy-brown. They feed on roots and the surface of the stems at about soil level, sometimes completely encircling the stem and causing the plant to die. Less extensive damage causes slow growth and a tendency to wilt during sunny weather, despite adequate soil moisture. When an affected plant is dug up, one or more cutworms can usually be found close to the soil surface. Plants can be given some protection against cutworms by treating the surrounding soil with chlorpyrifos granules. Fully grown caterpillars are, however, difficult to control with insecticides, and if plants do succumb it is worth while searching for the cutworm and destroying it before it moves on to another plant.

Stems with visible fungal growth
Pansy rust (*Puccinia violae*) causes distortion of stems, which become swollen and bear pustules producing first yellow-orange spores and, later, brown spores. Destroy affected plants as soon as symptoms are seen.

Stems galled
Leafy gall (*Corynebacterium fascians*) shows as a mass of aborted flattened shoots at ground level, particularly on antirrhinums and sweet peas. Destroy affected plants, and grow antirrhinums and sweet peas on a fresh site the following year.

Flowers lacking
Bud drop of sweet peas is a common trouble

and is caused by either cold night temperatures or too dry soil conditions. Nothing can be done to prevent this trouble when the nights are cold at flowering time, but in dry seasons water before the soil dries out completely, and mulch to conserve moisture.

Flowers rotting
Gray mold (*Botrytis cinerea*) causes rotting of sunflower and forget-me-not flowers in wet weather. Affected tissues become covered with a gray velvety mold. It can also cause spotting and rotting of sweet pea flowers. Cut off and burn affected flowers. In severe attacks spray with benomyl or thiophanate-methyl.
Petal blight (*Itersonilia perplexans*) can affect cornflowers, causing small oval water-soaked spots on the outer florets. In wet seasons it may spread rapidly and spoil the blooms. Cut off and burn affected plants. In gardens where this disease is known to occur, grow cornflowers well away from chrysanthemums and dahlias, since these can also become infected. Spray or dust with zineb before the buds open, repeating every week for as long as wet or humid conditions persist.

Flowers abnormal
Viruses such as cucumber mosaic and tomato spotted wilt can cause flowers to be small and misshapen, and sometimes spotted or streaked. Turnip mosaic virus causes white or yellow striping of honesty, stock and wallflower blooms, and various viruses cause flecking, mottling or striping of sweet pea flowers. Destroy affected plants and control aphids since they transmit these viruses.

Roots galled
Club root (*Plasmodiophora brassicae*) shows as irregular swellings on the roots of stocks and wallflowers, resulting in stunting of the plants. Raise seedlings in sterile compost if this disease has been troublesome in previous years. Grow susceptible plants on a fresh site each year where no cruciferous plant has been grown, in well drained soil which has been limed. Adjust the soil pH to 7 with hydrated lime and apply quintozene to the planting bed. Burn diseased plants at the end of the season.

Herbaceous plants 1

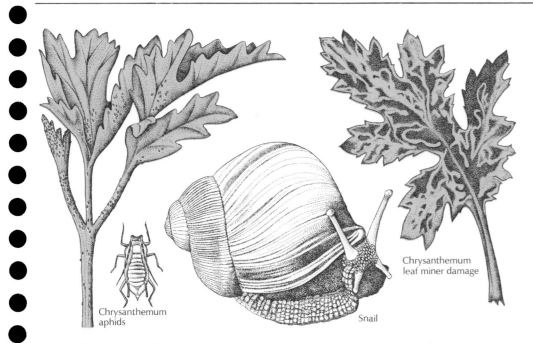

Chrysanthemum aphids

Snail

Chrysanthemum leaf miner damage

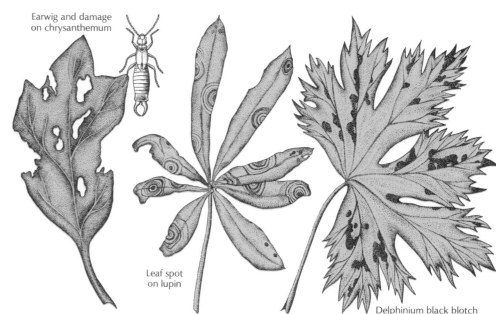

Earwig and damage on chrysanthemum

Leaf spot on lupin

Delphinium black blotch

Leaves with pests visible

Aphids (many species) suck sap from the leaves and stems, causing various symptoms such as stunted growth, poor flowering and curled leaves; these may also become soiled with honeydew upon which a black sooty mold may grow. Aphids may also spread virus diseases. Control them by spraying the plants throughly with pirimicarb, dimethoate or pirimiphos-methyl.

Leaves eaten

Slugs, snails and caterpillars cause holes to appear in the leaves. Slug and snail damage can be recognized by the silvery slime trail that they leave on the foliage. Control them by scattering slug pellets among the plants when the weather is warm and damp, since they usually come out to feed at this time. If caterpillars are causing the damage they can usually be seen on the plants, although it may be necessary to examine them by torchlight since many are nocturnal in their feeding habits. If the plants are too heavily infested for hand-picking, spray them with fenitrothion or derris.

Earwigs (*Forficula auricularia*) eat the soft parts of the leaves until only the network of veins remains. Earwigs can be very abundant in some years, and may attack a wide range of plants. Since they feed at night, an inspection by torchlight is necessary to find out the extent of an infestation. If the infestation is heavy spray the plants with either fenitrothion or trichlorphon at dusk on warm, still evenings.

Leaves mined

Leaf miners tunnel between the upper and lower surfaces of leaves causing a white or brown discolored area where the internal tissues have been eaten away. The shape of the mine depends upon the species of leaf miner causing it. It may be a narrow, meandering tunnel, as seen with the chrysanthemum leaf miner (*Phytomyza syngenesiae*) or it may be a large, irregular blotch mine, as shown by the delphinium leaf miner (*Phytomyza aconiti*). If a mined leaf is held up to the light it is usually possible to see either the maggot or its pupa inside the leaf. Control leaf miners by spraying the plants with either pirimiphos-methyl or trichlorphon as soon as damage is seen. Heavily infested plants may require two or three applications at ten day intervals.

Leaves discolored

Faulty root action due to adverse soil conditions can result in the foliage turning yellow, purple or brown. The most susceptible plants are Michaelmas daisies, chrysanthemums, phlox and romneya. Improve the soil if it is dry or waterlogged (see pages 5–7). Spraying with a foliar feed may help to restore the plant's color.

Magnesium deficiency can develop on chrysanthemums fed with high potash fertilizers. Yellow-orange blotches form between the veins, starting on the lower leaves and spreading upwards. Spray affected plants with a solution of magnesium sulfate made at the rate of $\frac{1}{2}$ lb in 3 gal of water, to which is added a spreader, repeating once or twice at two week intervals.

Leaves spotted

Leaf spot may be caused by many different fungi. Most are specific to one type of host and do not spread to other plants. In general the spots are brown or black and may coalesce until the whole leaf surface is covered. Severely affected leaves may wither and fall prematurely. On some spots, black pinpoint-sized fruiting bodies of the fungus appear and produce spores by which the disease spreads. Remove and burn affected parts as soon as the symptoms are seen. Spray with maneb, zineb or mancozeb, and repeat two or three times at two week intervals if further leaf spotting occurs. Note that waterlilies should not be sprayed with any of the above fungicides. Check whether the soil conditions around severely diseased plants are too wet or too dry, and carry out any necessary remedial measures. Spray all plants with a foliar feed to encourage vigor.

Black blotch (*Pseudomonas delphinii*) causes large black blotches to appear on the leaves of delphiniums. These may spread to the stems and even the flowers. This is a difficult disease to control but where it is known to be troublesome spray the soil surface and the young plants with a copper fungicide and repeat the treatment two or three times at two week intervals as necessary. Spraying should be done soon after the shoots appear above the soil.

Viruses such as arabis mosaic, cucumber mosaic and tomato spotted wilt can cause yellow concentric rings, ring and line patterns or a general mottling of the leaves of herbaceous plants. Paeonies, lupins, hostas

Herbaceous plants 2

Meconopsis
Nepeta
Nymphaea (waterlily)
Paeonia
Papaver
Pelargonium; zonal types
Pulmonaria
Pyrethrum
Romneya

Saxifraga
Sedum
Sempervivum
Sidalcea
Solidago
Trollius
Verbascum
Vinca
Viola (violet)

Perennial species of the
following genera:
Alyssum
Arabis
Campanula
Euphorbia
Malva
Phlox
Primula

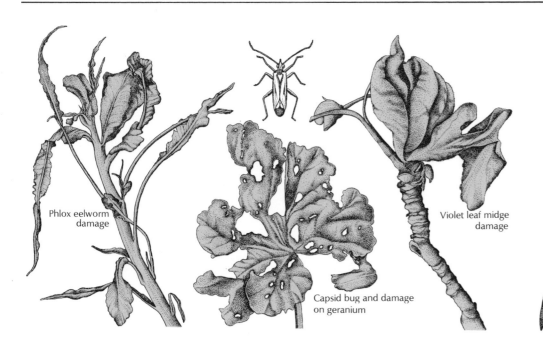

Phlox eelworm
damage

Violet leaf midge
damage

Capsid bug and damage
on geranium

Powdery mildew
on Michaelmas daisy

Hypericum rust

White blister on honesty

and delphiniums are particularly susceptible. The leaves may also be distorted and the plants stunted. On lupins cucumber mosaic can cause a progressive browning of the leaf stalks and stems until the whole plant is dead. Destroy any plant showing these symptoms.

Leaves blotched
Leaf and bud eelworms (*Aphelenchoides ritzema-bosi* and *A. fragariae*) are microscopic, worm-like creatures that live inside the leaves of many herbaceous plants, causing affected parts to turn brown or black. The spread of eelworms inside the leaves is restricted by the larger leaf veins, which sharply divide the healthy green parts from the infested areas. Chrysanthemums are particularly susceptible but paeony, pyrethrum, doronicum, Japanese anemone and many others may also be attacked. Usually the lower leaves are affected first, but symptoms later spread up the plants, particularly during wet weather when the eelworms emerge from the leaves and spread rapidly in the water film that covers the foliage. There are no chemicals available to amateur gardeners for controlling these pests and badly infested plants should be removed and burned.

Leaves distorted
Capsid bugs (*Lygocoris pabulinus* and *Lygus rugulipennis*) suck sap from the buds and young leaves. They have a toxic saliva that kills plant cells around the feeding site. Later, when affected leaves expand, these dead areas tear into many small holes, giving the foliage a tattered and distorted appearance. Damage appears first in late May or June and may continue until late summer. Check capsid bugs by spraying with dimethoate, formothion or fenitrothion as soon as damage is seen.
Violet leaf midge (*Dasineura affinis*) is a tiny gnat-like fly that lays its eggs among the folds of developing violet leaves. These hatch into orange-white maggots that grow up to $\frac{1}{8}$ in long and feed on the leaves. Their presence causes the leaves to become greatly thickened and tightly curled. If most of the leaves are damaged the plant may be killed, but violets readily seed themselves and make replacements. With small numbers of plants it is possible to check the midge by picking off and burning affected leaves. Alternatively spray the plants with dimethoate or pirimiphos-methyl when signs of recent leaf damage are seen.

Leaves and stems distorted
Phlox eelworms (*Ditylenchus dipsaci*) are minute, worm-like animals that live inside the leaves and stems of phlox. The stems thicken, and become twisted and split, especially at the base, and growth is stunted. Leaves at the shoot tips become distorted and reduced in width, sometimes to the extent of being little more than the midrib vein. Badly infested plants fail to flower and each year become progressively weaker. None of the chemicals available to amateur gardeners controls eelworms, and infested plants should be burned. Other plants that can be attacked by phlox eelworm include aubrieta, gilia, gypsophila, oenothera, *Collomia biflora*, sweet william and various weeds. These host plants should not be grown on infested soil for at least two years.

Leaves with visible fungal growth
Powdery mildews can affect a wide range of plants. A white powdery coating develops on the leaves and sometimes the stems and even the flowers. Since plants are more susceptible to these diseases if the soil is too dry, mulch to conserve moisture, and water in dry periods. At the first signs of trouble spray with benomyl, dinocap or thiophanate-methyl, repeating treatment as necessary. Note that dinocap may injure certain varieties of chrysanthemum. At the end of the season cut off and burn all diseased stems and debris in order to eliminate the over-wintering stages of the fungi.
Rust is caused by a variety of fungi. On artemisia, periwinkle, zonal pelargoniums, chrysanthemum and border carnations, dark brown masses of spores develop on the leaves. On other plants the pustules are usually yellow or orange, but later in the season may produce brown spores. In slight attacks remove affected leaves and spray with maneb, zineb or mancozeb repeating treatment at two week intervals as necessary. At the end of the season burn all infected shoots. Where the disease appears every year on a plant, destroy it because the fungus is probably systemic within all its tissues and no cure is possible.
White blister (*Albugo candida*) causes white blisters or swellings containing white powdery masses of spores on the leaves and stems of alyssum, arabis and honesty. The pustules may be single or in concentric rings. Remove and burn affected leaves and stems.

Herbaceous plants 3

Root rot on chrysanthemum

Smut on trollius

Michaelmas daisy wilt

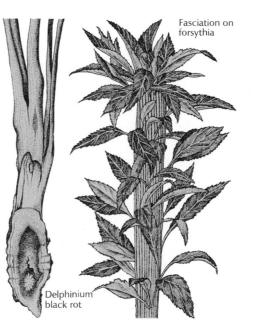

Fasciation on forsythia

Delphinium black rot

Smut is caused by *Urocystis anemones* on trollius, and by *Urocystis violae* on violets. The leaves and stems develop elongated swellings or blisters which burst to discharge black dusty spores. Destroy all infected plants.

Leaves and stems rotting
Leaf and stem rot of carnations and pinks is caused by *Alternaria* spp and *Heteropatella valtellinensis*. White or brown-yellow spots with dark margins, or gray spots, may appear depending on which fungus is causing the trouble. These spots develop on the leaves and stems, and may even affect the flower buds. Affected tissues eventually rot. Remove diseased parts and spray the remains of the plants with a copper fungicide.

Stems dying or wilting
Paeony wilt (*Botrytis paeoniae*) causes the bases of paeony shoots to turn brown and wither, resulting in wilting and, finally, death of affected shoots. A gray furry growth of fungus develops on the diseased tissues but is not easily noticed. Brown angular patches may develop on the leaves, usually at the tips, and even the flower buds may turn brown and die. Cut out affected shoots to

well below ground level and burn them. Spray the crowns of the plants with benomyl or dichlofluanid soon after the leaves emerge, repeating at two week intervals until flowering and again after flowering, if necessary. Since the disease is favored by damp conditions, thin very dense clumps and do not apply nitrogenous fertilizers since this encourages soft growth.

Michaelmas daisy wilt (*Phialophora asteris*) affects all asters with the exception of varieties of *Aster novae-angliae*. Only certain shoots in a clump are affected; these show browning and withering of the leaves starting at the base of the stem and spreading upwards. The leaves do not fall even though the shoots wilt and die. An affected plant may die after two or three years. Where there is ample stock, destroy affected plants and propagate only from healthy ones. When a valued plant is diseased, propagate by taking cuttings only one or two inches long from the tips of the suckers, and root them as cuttings. Plant them on a fresh site because the disease is soil-borne.

Verticillium wilt (*Verticillium* spp) causes the leaves of chrysanthemums to turn yellow and wilt from the base upwards. Infected stems

will show brown streaks in the inner tissues if they are cut longitudinally. Dig up and burn affected plants. Do not replant the infected area with chrysanthemums unless the soil is sterilized with Vapam at least three weeks before replanting. Alternatively, plant chrysanthemum varieties that are certified to be verticillium resistant.

Fusarium wilt of dianthus, especially pinks, is caused by the soil-borne fungus *Fusarium oxysporum* f *dianthi*. The leaves become yellow and wither, and the plant wilts. Brown longitudinal streaks develop in the inner tissues. Destroy affected plants, and grow pinks and other dianthus on a fresh site for the next few years. Treat the soil with Vapam at least three weeks before replanting with new stock. In slight attacks it may be possible to prevent spread of the disease by drenching plants with a solution of benomyl or thiophanate-methyl, repeating two weeks later.

Root rot is usually caused by the soil-borne fungi *Thielaviopsis basicola* and *Rhizoctonia solani*. Most types of plants can be affected but Michaelmas daisies, chrysanthemums, delphiniums, primulas, gentians and lupins are the most susceptible. The roots and crown tissues become black and rotten, the

leaves become discolored from the tips of the shoots downwards and the shoots die back. Lift plants showing these symptoms, and destroy any that are nearly dead. In less severe cases cut out the dead parts from the crowns, remove dead roots and those with black or brown patches and dust the remains of the plants with quintozene. Replant on a fresh site and spray developing foliage with a foliar feed to encourage vigor.

Black rot affects delphiniums. The cause of this disease is not known, though it has similar symptoms to root rot (see above), but the inner tissues of the crown also become black, and are often hollow, as can be seen if the crown is cut longitudinally. The dead stems break off easily at ground level. Destroy affected plants, preferably by burning them, and grow delphiniums on a fresh site the following year.

Brown core (*Phytophthora primulae*) can attack perennial primulas. For symptoms and treatment, see page 69.

Honey fungus (*Armillaria mellea*) can attack paeonies, delphiniums and lupins. Affected plants usually die rapidly but the dead leaves do not fall. White streaks of fungal growth develop within the dead tissues of the roots and crown, which may suffer from a mushy rot. The fungus spreads through the soil by means of dark brown root-like structures. These may be found attached to the rotting roots. Dig up affected plants, together with all the roots, and burn them. Grow annuals on the site for the next few years. If the infected bed is vacant sterilize with 2 per cent formalin. Try to trace the source of infection, and dig up and burn all woody debris.

Faulty root action due to adverse soil conditions can cause shoots of affected plants to wilt and even die. For treatment see under Leaves discolored, page 71.

Stems distorted
Fasciation, or flattening of shoots, may be due to an early injury to the growing point caused by frost, an early insect attack, slug damage or even mechanical injury caused by hoeing. Most plants can be affected but the most susceptible are delphiniums, euphorbias and primulas. No treatment is necessary since the problem is not serious.

Herbaceous plants 4

Stems with frothy liquid

Cuckoo spit is a white frothy liquid secreted by nymphs of sap-feeding insects known as froghoppers. It occurs in May or June on the stems of many plants. If the liquid is gently blown away, a creamy-white nymph will be exposed. They are not serious pests since little harm is caused, although if the nymph is feeding at a shoot tip it may cause some distortion of the new growth. Either pick the nymphs off by hand or spray plants with a systemic insecticide such as dimethoate or formothion.

Stems splitting

Splitting of stems, particularly of phlox and pelargoniums, is usually due to irregular growth and is caused most frequently by heavy rain or watering following a dry spell. The splitting usually occurs towards the base of the stem, which opens out to reveal the inner tissues. Affected parts may also twist and severely split stems may die. Cut off affected stems, and prevent this trouble by mulching to conserve moisture and by keeping the soil moist in dry periods.

Stems galled

Leafy gall (*Corynebacterium fascians*) is found most commonly on chrysanthemums, phlox and heuchera, and shows as a mass of aborted and fasciated shoots at ground level. Destroy affected plants and do not grow susceptible hosts in the infected area.

Stems stunted

Viruses such as cucumber mosaic, arabis mosaic and tomato spotted wilt may cause stunting of plants, particularly delphiniums and lupins. For further symptoms and treatment, see under Leaves spotted, page 71.

Stems or crowns rotting

Gray mold (*Botrytis cinerea*) affects mainly euphorbias, and may cause rotting of stems that have been injured by frost. Affected tissues become covered with a gray velvety growth of fungus. Cut out affected shoots and spray with benomyl.
Crown rot of lewisias and sedums is a physiological disorder caused by excessive moisture in wet seasons. Destroy severely

affected plants. In slight attacks cut out any rotting tissues and dust the crowns with dry copper fungicide.
Nymphaea crown rot of waterlilies is caused by species of the fungi *Pythium* and *Phytophthora*. The leaves and stems are affected by a black slimy rot. Destroy affected plants and protect other plants in the pond by treating the water with a solution of copper sulfate. For every 1,500 cu ft of water place $2\frac{1}{2}$ oz copper sulfate into a cloth bag and pull it backwards and forwards across the pond until it has dissolved. A solution of copper sulfate at this strength should not harm any fish except trout.

Flower buds not opening

Poor flowering of paeonies can be due to a physiological disorder that causes the flower buds to remain small and hard. This trouble can be caused by frost damage, the soil being too dry, malnutrition, planting too deeply or root disturbance. Prevent it by planting paeonies with the buds at or just above soil level, keep them well mulched, preferably with farmyard manure which will provide all the food materials they need, and water in dry periods. Weed carefully around paeonies so as not to injure the roots and crowns. Protect buds with newspaper or old net curtains if a severe frost is forecast.
Paeony wilt (*Botrytis paeoniae*) may prevent the buds from opening. Affected buds turn brown and die, and become covered with a gray velvety fungal growth. For control, see under Stems dying or wilting, page 73.

Flower stalks wilting

Pedicel necrosis is a physiological disorder that shows as a blackening of the top 2 in of the flower stalk. The stalk usually topples at the affected area and the flower buds turn brown and fail to open. This trouble occurs most frequently on poppies and pyrethrums. It is caused by the plant having insufficient vigor to produce the full complement of flowers, either as a result of malnutrition or because of a check in growth due to sudden drying out of the soil. Cut off affected flower stalks and, in late summer or the following spring, apply a dressing of sulfate of potash at $\frac{1}{2}$ oz per square yard to help the plant

Paeony wilt

Gray mold on chrysanthemum

Delphinium mosaic

Leafy gall on chrysanthemum

Cuckoo spit, caused by froghoppers

Herbaceous plants 5

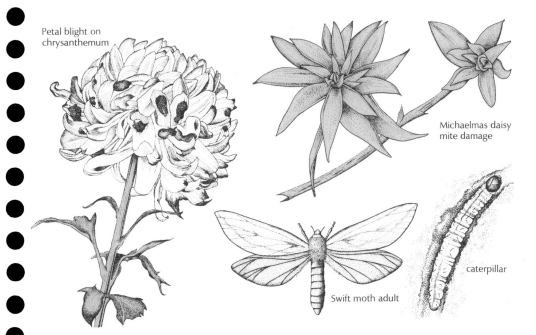

Petal blight on chrysanthemum

Michaelmas daisy mite damage

Swift moth adult

caterpillar

VIRUS DISEASES OF CHRYSANTHEMUMS

Of the many different virus diseases that may affect chrysanthemums, the most important is aspermy, which affects the blooms. The florets may be twisted, quilled, tubular or have uneven length, so that the blooms appear distorted and small. The flower color is broken by pale streaks or flecks. This disease is spread by aphids so control these pests by spraying with a suitable insecticide. Stunt is another serious virus disease in which the plants are severely stunted and flower abnormally early. It is spread by contact between diseased and healthy plants, and when plants are handled. Leaf mottling, flower distortion and stunting may also be caused by other viruses. In all cases destroy affected plants. Afterwards, thoroughly wash hands, and any equipment that may have been used, in hot soapy water.

Aspermy

overcome the problem. In addition mulch the plant the following season to conserve moisture, and ensure that the soil does not dry out completely.

Flowers eaten
Earwigs (*Forficula auricularia*) damage the flowers of chrysanthemum and similar types of bloom. Control them by spraying with either fenitrothion or trichlorphon on warm, still evenings when damage is noticed. Alternatively, trap them by placing rolls of corrugated cardboard, or clay pots loosely stuffed with straw, among the plants. The earwigs hide there during the day and may be picked out and destroyed.

Flowers distorted
Viruses such as arabis mosaic, cucumber mosaic and tomato spotted wilt can cause distorted or poor quality flowers on delphiniums. Destroy affected plants.

Flowers discolored
Michaelmas daisy mites (*Tarsonemus pallidus*) are minute creamy-white mites that infest the flower buds and leaf sheaths. The stems develop a rough brown scarring where the mites have been feeding, and growth is stunted. Flowers fail to develop normally and rosettes of small green leaves are produced instead of petals. There are no chemicals available to the amateur gardener that are effective against this type of mite. Heavily infested plants produce so few normal flowers that they are not worth keeping and should be burned. Michaelmas daisies of the *Aster novi-belgii* type are often badly damaged but *Aster novae-angliae* and *Aster amellus* can usually flower normally despite being infested with the mite.

Black blotch of delphiniums (*Pseudomonas delphinii*) causes black blotches on the flowers. For control measures see under Leaves spotted, page 71.

Leaf spot of hellebores is caused by the fungus *Coniothyrium hellebori*. In severe attacks it can cause black blotches on the flower buds and petals. Cut off all affected parts and spray with a copper fungicide, repeating at two week intervals until the plant comes into flower.

Viruses of many types can cause white flecks or streaking of flowers particularly on carnations, delphiniums and lupins. Destroy affected plants.

Mycoplasma, an organism intermediate between a virus and a bacterium, can cause helenium, aster and primula flowers to become green. Destroy affected plants.

Flowers rotting
Gray mold (*Botrytis cinerea*) may cause rotting of flowers in a very wet season, affected tissues being covered with a gray velvety mold. Cut off rotting blooms. If chrysanthemums are severely attacked spray plants with benomyl or thiophanate-methyl, repeating at two week intervals until symptoms disappear.

Petal blight (*Itersonilia perplexans*) attacks chrysanthemums, causing the outer florets to develop small oval water-soaked spots. In wet weather the disease spreads rapidly until the whole blossom becomes brown and rotten. Gray mold then frequently sets in. In damp weather a dull white bloom develops on the affected tissues due to the production of great numbers of spores by which the disease spreads to adjacent flowers. Pick off and burn affected florets, or whole blooms if necessary, and spray with zineb. Where this disease is known to be troublesome commence spraying with zineb when the flower

buds first show color and repeat at five to seven day intervals until the plant comes into flower.

Roots damaged
Swift moth caterpillars (*Hepialus* spp) may occur in some numbers among the roots of herbaceous perennials such as Michaelmas daisy, phlox, paeony, delphinium and chrysanthemum. They grow to 2 in long and have red-brown heads and white bodies that are sparsely covered with dark hairs. If growth is being impaired drench the soil with trichlorphon. When dividing and replanting herbaceous perennials look for and remove any swift moth caterpillars that may be hidden among the roots.

Ants (various species) often build their nests among the roots of herbaceous perennials. Little direct damage is caused but their extensive tunneling may loosen soil around the roots and make the plants more susceptible to dry conditions. When ants construct their nest a fine soil is deposited on the surface which may almost bury dianthus, saxifrage and other low-growing plants. Ants are such abundant garden insects that it is not possible to eliminate their nests.

Roses 1

Leaves with pests visible

Aphids (various species) may form dense clusters on the younger leaves, stems and flower buds between late April and the end of summer. Heavy infestations weaken plants, which become soiled with the aphids' sticky honeydew and the black sooty mold that grows on it. Control them by spraying with dimethoate, formothion, menazon or pirimicarb at intervals during the summer.

Leaves skeletonized

Rose slugworm larvae (*Endelomyia aethiops*) are pale green caterpillars up to $\frac{5}{8}$ in long with light brown heads. They feed mainly on the undersides of leaves where they graze away the lower surface until only the larger leaf veins and the upper, outermost layer of leaf remain. This layer dries up and turns white or brown. There are two generations, with caterpillars occurring in mid-May to June and again in July to August. The second generation is generally the more numerous and damaging. Control them by spraying with derris, fenitrothion, pirimiphos-methyl or trichlorphon.

Leaves with holes

Caterpillars (various species) may eat irregularly shaped holes in the foliage. When damage is seen, spray roses with the insecticides recommended for rose slugworm.

Leaf-cutting bees (*Megachile* spp) also make holes in the foliage, but their damage can be distinguished readily from that caused by caterpillars by the regular shape and size of the missing portions. These are always taken from the leaf margin and have a smooth outline. The pieces are either nearly circular or lozenge-shaped, and are used as nesting material by the female bees. Leaf-cutting bees are useful pollinators, so light attacks should be tolerated. If controls are necessary to prevent too much damage, keep a watch on the affected plants and swat the bee when it returns to remove another piece of leaf.

Leaves discolored

Faulty root action due to drought, waterlogging or general malnutrition can cause the leaves to turn yellow or brown. Prevent such troubles by careful planting and correct cultural treatment including feeding, mulching and watering as necessary. If the soil is prone to waterlogging, carry out appropriate remedial measures (see pages 5–7). It is often possible to restore vigor to affected plants by spraying with a foliar feed during the growing season.

Magnesium deficiency symptoms are common on roses, especially if growing on sandy soils in wet seasons or following heavy applications of potash fertilizers. It shows first on the old leaves as yellow or orange blotches between the veins, followed by browning of the affected tissues. The trouble then spreads upwards. Spray affected plants with $\frac{1}{2}$ lb of magnesium sulfate in 3 gal of water to which is added a spreader. Repeat once or twice at two week intervals. Prevent this trouble by using a rose fertilizer that contains magnesium.

Viruses such as prunus necrotic ring spot and arabis mosaic produce yellow spots and blotches, especially on the younger leaves, or a conspicuous yellow banding of the veins. Since there is no method of curing an affected bush, no treatment is necessary and flowering is not affected. The trouble usually occurs as a result of infected rootstocks, so buy roses from reputable firms.

Cold weather can cause a pink or purple discoloration, or sometimes brown blotches, on young soft leaves. In severe cases affected bushes may lose vigor, in which case spray with a foliar feed.

Rose leafhoppers (*Edwardsiana rosae*) are pale yellow insects about $\frac{1}{8}$ in long, which suck sap from the undersides of the leaves. When disturbed the adults readily jump off the leaves, although their creamy-white nymphs are much less active and remain on the plant. Where they have been feeding, they cause a white or pale green mottled discoloration of the upper leaf surface and, in heavy infestations, nearly all the green color may be lost. Roses growing in warm positions, such as against a wall, are especially susceptible to attack. Control them with systemic insecticides such as dimethoate or formothion.

Red spider mites (*Tetranychus urticae*) are mainly a greenhouse problem, but they may also attack outdoor plants during a warm summer. Wall-trained roses and those that

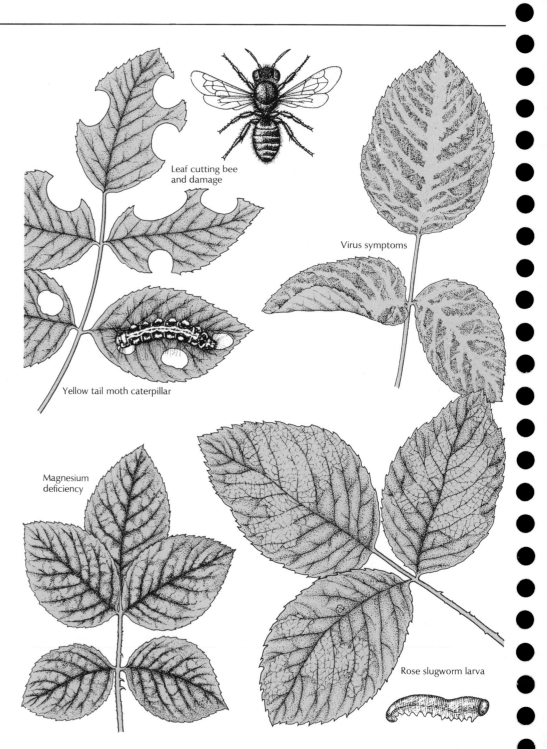

Leaf cutting bee and damage

Virus symptoms

Yellow tail moth caterpillar

Magnesium deficiency

Rose slugworm larva

Roses 2

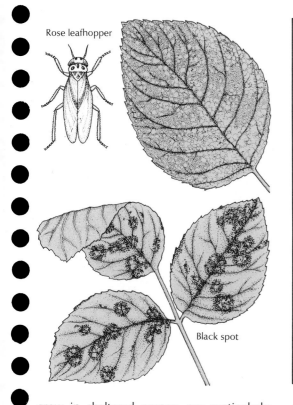

Rose leafhopper

Black spot

grow in sheltered corners are particularly susceptible. The mites are only just visible to the naked eye and are yellow-green with dark markings. They infest the undersides of leaves where they feed by sucking sap. This causes a fine mottled discoloration of the upper leaf surface, with the leaves turning dull green at first. Later they may turn yellow and fall off. Red spider mites are not easy to control, especially if a heavy infestation has been allowed to develop. Spray plants with dimethoate, formothion or malathion on three occasions at seven day intervals as soon as damage is noticed.

Leaves spotted

Black spot (*Diplocarpon rosae*) is probably the most common disease of roses. Circular dark brown or black spots up to $\frac{1}{2}$ in in diameter develop on the leaves. They are surrounded by yellowing tissues, and as they increase in size the whole leaf becomes discolored and falls prematurely. In severe cases complete

GENERAL FUNGAL GROWTH

Powdery mildew (*Sphaerotheca pannosa* f *pannosa*) shows as a white floury coating of fungus spores on the leaves, shoots, buds and, occasionally, the open blooms. It is more severe on plants that are dry at the roots—wall climbers are particularly susceptible since the soil there can become very dry. Mulch plants well to conserve moisture, and water in dry periods. The fungus over-winters as a gray felt-like growth on the shoots, particularly around and on the thorns. Cut out affected shoots when pruning and spray with dinocap, bupirimate, triforine, benomyl, carbendazim or thiophanate-methyl. However, the fungus could develop tolerance to the last three fungicides if they are used too frequently. Some varieties, such as 'Dorothy Perkins' and 'Frensham', are very susceptible and should not be grown where mildew is common. Instead, grow resistant varieties such as 'Ena Harkness', 'Peace' and 'Queen Elizabeth'—however, even these may be susceptible in some localities.

defoliation of the bush can occur. Sometimes the diseased patches are small and diffuse, particularly on *Rosa* species, and in this form black spot is less easily recognized. The fungus over-winters on the fallen leaves and, rarely, in scabby lesions on the stems. Rake up and burn fallen leaves. Protect the bushes by spraying them with captan, dichlofluanid, maneb or zineb immediately after pruning, and repeat at two week intervals throughout the season. Benomyl and thiophanate-methyl will also control black spot but regular use of these chemicals could lead to the development of tolerant strains of the fungus. If black spot persists in spite of spraying regularly, the bush may be lacking in vigor, in which case spray with a foliar feed during the growing season and improve the growing conditions (see also pages 5–7). Should the disease still persist grow varieties that are less susceptible to black spot, such as 'Ena Harkness', 'Sunset', 'Pink Peace', 'Sutter's Gold', 'Allgold', 'Iceberg' and 'Queen

Elizabeth'. The susceptibility of varieties differs from district to district and even those listed above may be attacked by this disease in certain localities.

Leaves distorted

Hormone weedkiller damage, caused by the misuse of selective weedkillers, is a very common trouble on roses. Affected shoots and leaf stalks twist spirally and bear distorted narrow, twisted and often cupped leaves, which show parallel veining. Cut off affected shoots. Prevent this trouble by the careful use of hormone weedkillers, which should be applied only on a still day using equipment kept specifically for their use. Do not mulch with grass cuttings taken from a lawn recently treated with a hormone weedkiller.

Capsid bugs (mainly *Lygocoris pabulinus*) are pale green insects, about $\frac{1}{4}$ in long, which suck sap from the buds and young leaves. When these young leaves expand to their full size, they tear into many small holes and may be distorted. By the time this damage is recognized it is usually too late to take control measures as the bugs will have moved on to some other plant. Capsid bugs are controlled by systemic insecticides, and the routine use of compounds such as dimethoate or formothion against aphids will also control capsids. In gardens where capsids are a persistent problem spray roses with one of these insecticides in early May, even if aphids are not present.

Leaves rolled

Leaf-rolling rose sawfly (*Blennocampa pusilla*) is a small, black, fly-like insect that lays its eggs in rose leaves in mid-May or June. Affected leaves become tightly rolled up within 24 hours due to chemicals secreted by the female sawflies. After a week or so, the eggs hatch into pale green caterpillars which feed on the rolled leaves. There is only one generation a year, and when fully fed in June or July the caterpillars go into the soil to pupate. Deal with light attacks by picking off and burning the affected leaves before the larval stage is completed. Extensive infestations will need spraying with pirimiphos-methyl or trichlorphon in mid-June.

Leaves with visible fungal growth

Rust may be caused by any of several species of *Phragmidium*. In spring bright orange patches up to 1in long may develop on the stems and leaf stalks. This stage of the fungus, however, is not very common. In summer, yellow spots show on the upper leaf surfaces and orange powdery pustules develop on their undersides. Later the pustules turn brown then black, and the leaves dry up, become brittle and fall prematurely. In early or severe attacks affected bushes may be so weakened that they die-back. Destroy such bushes, and rake up and burn all affected leaves. Cut out the spring stage of the disease when present. Maneb, zineb and mancozeb fungicides will give some control of rust, but the most effective fungicide is oxycarboxin. If using oxycarboxin, ensure that it is labeled for greenhouse use only. As the disease is worse on weak bushes, spray with a foliar feed during the growing season, mulch them well to conserve moisture, and water them before the soil dries out completely.

Stems with pests visible

Aphids may attack the stems and leaves. For symptoms and treatment, see under Leaves with pests visible, page 76.

Scale insects, such as brown scale (*Parthenolecanium corni*) and scurfy scale (*Chionaspis furfura*) may attack roses. The former have convex, oval, red-brown shells up to $\frac{1}{4}$ in long, and occur mainly on roses growing in warm situations such as against a wall. Scurfy scale encrusts the stems of species roses rather than the hybrid tea types. The shells of this scale insect are gray white, mainly flat and circular in shape and about $\frac{1}{10}$ in in diameter. They are often so closely packed together that it is difficult to see individual scales. Control scale insects by spraying with malathion in early July for brown scale and in mid-September for scurfy scale. Dead scales may remain attached to the stems for some years but new growth should remain free of scales after treatment.

Stems discolored

Frost damage can result in the development of purple patches on the stems. In most cases the damage is only superficial, and no

Roses 3

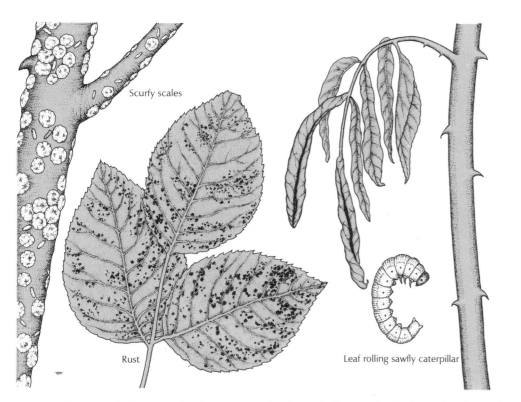

Scurfy scales

Rust

Leaf rolling sawfly caterpillar

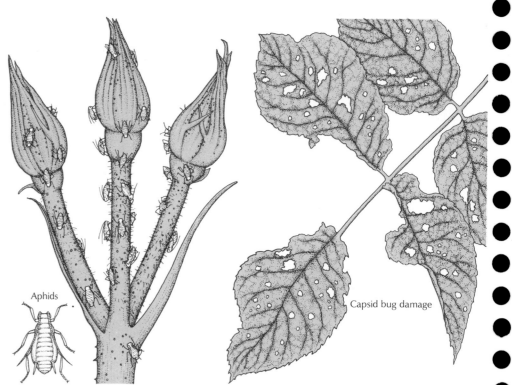

Aphids

Capsid bug damage

treatment is needed. Occasionally, however, damaged shoots suffer from a secondary infection of gray mold (see below).

Stems galled
Crown gall (*Agrobacterium tumefaciens*) may show as a chain of small galls on the shoots, or less frequently as a walnut-sized gall at ground level. Burn severely affected bushes and do not plant another rose in the same position. If another shrub is to be planted dip the roots in a copper fungicide before planting and try to avoid injuring the roots. Improve the drainage where necessary.

Stems blind
Blindness of shoots, in which the tip aborts, is usually caused by frost damage or large differences between the day and night temperatures. Cut off affected shoots to encourage the development of side-shoots lower down. If further trouble occurs, young growth may be too soft from excess nitrogen

in the soil. If so, rake in ½ oz of sulfate of potash per square yard.

Stems dying back
Faulty root action can be caused by drought, waterlogging, malnutrition or poor planting —that is, planting the bushes too deeply or not spreading out the roots sufficiently during planting. Carry out any necessary improvements to the soil (see pages 5–7). If poor planting is the cause of the trouble, lift the affected bushes in the fall and replant them more carefully.

Canker is caused by the fungi *Griphosphaeria corticola* and *Leptosphaeria coniothyrium*. In most cases the shoots merely die back and show brown patches on which may be seen pinpoint-sized eruptions. In severe cases, particularly if *Leptosphaeria coniothyrium* is responsible, a canker develops at ground level. The stem becomes swollen and the bark roughened and cracked. Such cankers are usually an inch or so in length and may

completely girdle a stem. Cut out all dying and cankered shoots. Where the tissues at the crown are decaying, remove all rotten wood and dust the crown with dry copper fungicides. In severe cases it may be necessary to lift an affected bush so that it can be replanted more carefully, making sure the graft union between stock and scion is not covered by soil since this may also encourage the fungi. If canker is troublesome, paint large pruning cuts with a fungicidal wound paint and spray the bushes at pruning time with a copper fungicide. Prevent infection by planting roses carefully in well prepared soil, and feed, mulch and water them in dry periods. Remove all dead and damaged wood as soon as it is seen. Finish pruning cuts cleanly and make each cut immediately above a bud and as close as possible to it.

Gray mold (*Botrytis cinerea*) can enter through wounds and cause die-back of the shoots. The affected tissues become covered

with a gray-brown furry growth of fungus. Cut out affected shoots about 2 in below the apparently dead tissues. Check whether the affected bush needs feeding, mulching, watering or replanting more carefully to encourage vigor, and carry out any necessary improvements.

Honey fungus (*Armillaria mellea*) frequently kills bushes, but it is not usually noticed until an affected plant dies suddenly. The fungus attacks from the soil and spreads as sheets of white fan-shaped growths beneath the bark of larger roots and the base of the stem. Dark brown root-like structures known as rhizomorphs develop on diseased roots and grow out through the soil to infect other plants. Remove dead and dying plants together with as many roots as possible. Sterilize the soil with 2 per cent formalin solution applied at the rate of 6 gal per square yard. Try to trace the source of the infection and destroy any old stumps and woody debris that are buried in the soil.

Roses 4

Vaporer moth caterpillar

Canker

Proliferation

Gray mold

Thrips damage (normal bud, left)

Gray mold following frost damage

Rose sickness is a form of replant disorder, the cause of which is not known. The trouble may occur when new roses replace old ones in the same soil. Affected bushes die back and appear sickly. The only way to prevent this trouble is to grow new roses on a fresh site, or to sterilize the soil with 2 per cent formalin.

Flower buds withering
Pedicel necrosis is a physiological disorder that shows as a blackening of the flower stalk for about 1 in just below the flower bud. The flower bud turns brown and fails to open. This trouble is caused by the plant having insufficient vigor to produce the full complement of flowers, either as a result of malnutrition or because of a check in growth due to a sudden drying out of the soil. Cut off affected flower stalks, but no other treatment can be carried out during the flowering period. If it is the first crop of flower buds to be affected, apply a foliar feed since this may enable it to flower normally later in the year. In late summer or the following spring apply a dressing of sulfate of potash at $\frac{1}{2}$ oz per square yard, and feed the rose with a complete fertilizer or special rose fertilizer in the spring and immediately after the first flowering. Mulch to conserve moisture, and water in dry periods. If the trouble persists poor growing conditions may be responsible, in which case see the section on physiological disorders, pages 5–7.

Dumpy bud is caused by a lack of vigor or a check in growth. The bud may not open, and if it does the sepals and petals are very short and give the flower a flattened dumpy appearance. Treat as for pedicel necrosis (see above).

Capping is caused by an injury to the petals, either from heavy rain or hail stones or as a result of a check in growth. The flower bud swells but does not open because the outer petals have stuck together. Buds affected by this trouble are frequently attacked by gray mold (see below). Treat as for pedicel necrosis (see above).

Flower buds rotting
Gray mold (*Botrytis cinerea*) sometimes causes the flower buds to turn brown and decay. In damp weather partially opened blooms may also be attacked. Affected tissues become covered with a gray fluffy growth of the fungus and should be cut off. If the disease is due to adverse weather conditions then no other treatment should be necessary. If affected blooms have a capped or dumpy appearance, treat as for pedicel necrosis (see above).

Buds and flowers eaten
Caterpillars of various species may eat the flowers and buds. Their damage prevents the blooms from developing normally. Control them by spraying roses with any of the insecticides derris, fenitrothion, pirimiphos-methyl or trichlorphon.

Buds and flowers discolored
Thrips (*Thrips fuscipennis*) are thin, black or yellow insects about $\frac{1}{10}$ in long which live between the petals, especially in partly open buds. They feed by sucking sap, causing brown streaks on the petals. Heavy infestations may result in distorted blooms. Damage can occur from May onwards and is more likely to be seen in hot dry summers. Control thrips by spraying with dimethoate, formothion or pirimiphos-methyl.

Flowers distorted
Proliferation is nearly always caused by early injury to the growing points when the flower buds are just beginning to develop. The main growing axis continues growing through the center of the flower so that one or more small flower buds develop within the center of the bloom. The buds generally remain small and hard. Very occasionally they may show some signs of colored petals but they do not open out into fully developed flowers. This condition is seen most frequently in old cabbage roses, Bourbon roses such as 'Mme Isaac Pereire' and 'Souvenir de la Malmaison', and the hybrid bush rose 'Felicia'. It can, however, occur on hybrid tea roses in a season when there are severe frosts at the time the flower buds are beginning to develop. Cut off affected flowers. Should the second crop of flowers also be affected and the trouble persist for several years, destroy the bush since the symptoms may then be caused by a virus infection.

Rhododendrons and azaleas 1

Leaves with pests visible

Rhododendron leafhoppers (*Graphocephala coccinea*) feed by sucking sap from the leaves during the summer. They cause no direct damage to rhododendrons but, as they lay their over-wintering eggs in the flower buds, they make incisions that provide entry sites for the spores of bud blast disease (see page 82). Adult leafhoppers are present on the plant from late July to October. They are $\frac{3}{8}$ in long and turquoise-green with orange-red markings on their wings and abdomen. The adults readily fly off the plant when disturbed, but their yellow-green nymphal stages are less active and may be found on the undersides of the leaves. If plants are affected by bud blast disease, spray with either fenitrothion or a systemic insecticide during August to October to control the leafhoppers.

Rhododendron whitefly and azalea whitefly (*Dialeurodes chittendeni* and *Pealius azaleae*) are tiny, white, moth-like insects that may occur in large numbers on the undersides of leaves during the summer. The adults and their flat, oval, pale green larvae feed by sucking sap and excrete a sugary substance known as honeydew. This makes the foliage sticky and allows the growth of a superficial black sooty mold. Control whitefly by spraying thoroughly the undersides of the leaves with a pyrethroid insecticide or pirimiphos-methyl. Several applications at about ten day intervals may be necessary since the immature stages are less susceptible to insecticides than the adults.

Leaves eaten

Black vine weevils (*Otiorhynchus sulcatus*) are active at night during the summer months. They are slow-moving, dark beetles about $\frac{3}{8}$ in long, and are more likely to be seen if the plants are examined by torchlight. They feed on the leaf margins, making irregularly shaped notches which give the foliage a ragged appearance. Low-growing branches are generally the most heavily damaged. Little real harm is likely to be caused to an established plant but, if control measures are required, dust or spray the foliage and soil surface with carbaryl at dusk when signs of recent damage are noticed.

Caterpillars of various moths feed on the leaves of rhododendrons. Their damage is distinguished from that caused by black vine weevils (see above) by the fact that it is more extensive and that it occurs in the center of the leaf as well as at the margins. Caterpillars rarely cause severe damage but, if many leaves are being eaten, it may be worth while

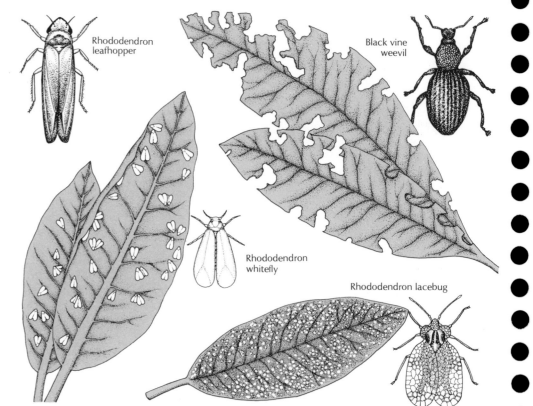

Rhododendron leafhopper

Black vine weevil

Rhododendron whitefly

Rhododendron lacebug

spraying the plant with pirimiphos-methyl, fenitrothion or trichlorphon.

Leaves discolored

Rhododendron lacebug (*Stephanitis rhododendri*) usually causes trouble on plants growing in exposed, sunny positions. Overwintering eggs are laid on the leaves, and hatch in May. When fully grown in July, the lacebugs are about $\frac{1}{4}$ in long and dark brown. The wings are carried flat on the insect's back, and the presence of numerous veins in the wings gives them a lace-like appearance. The adults and their nymphs live on the undersides of the leaves where they suck sap. This causes a yellow mottled discoloration to appear on the upper surface of the leaves, while the lower surface becomes a rusty-brown color. Control lacebugs by spraying the undersides of the leaves thoroughly with malathion, fenitrothion or pirimiphos-methyl.

Lime-induced chlorosis is due to alkaline soil conditions "locking up" iron and manganese, causing them to be unavailable to plants. Affected plants show yellowing between the veins and in severe cases the leaves at the tips of the shoots may be almost entirely pale yellow. Rhododendrons grow best in soils with a pH between 5.0 and 5.5, and where the pH is higher some chlorosis is likely to occur. Try to reduce the pH of the soil by digging in acidic materials such as peat, pulverized bark or crushed bracken, or use acidifying chemicals such as sulfur or aluminum sulfate. Do not use the latter chemical after planting. Continue to top-dress annually with organic acidic materials. Even with these treatments, in a soil of pH 7.5 or higher it will not be possible to reduce the pH to as low as 5.5. Therefore apply annually a proprietary product containing chelated compounds, or rake in fritted trace elements. Prevent this trouble by

---

FAULTY ROOT ACTION

The commonest cause of discoloration of foliage and die-back of shoots on rhododendrons is faulty root action. This may be due to growing them in too alkaline a soil (see Lime-induced chlorosis), the soil being too wet or dry, malnutrition or allowing the roots to dry out before planting. Poor planting may also be responsible, either because the shrub was planted too deeply or because the roots of pot-bound plants were not teased out gently during planting. Shrubs that have been poorly planted will be small and stunted, but they may not die back. Such shrubs may also fail to produce good flowers, either because no buds form, or the buds may fall before opening, or the flowers may only partly open. Too dry a soil

when flower buds are developing in late summer or early fall can also cause bud drop at flowering time.

Prevent such troubles by first checking that the soil is sufficiently acid before planting rhododendrons. Prepare the soil well by digging in plenty of decayed organic matter. Plant carefully, mulch well to conserve moisture, and water in dry periods before the soil dries out completely, particularly in late summer and early fall. Give a liquid feed when the shrubs are established satisfactorily. Feed mature shrubs annually by raking in a complete fertilizer at 3 oz per square yard. During the growing season spray shrubs that are weak or affected by drought or waterlogging with a foliar feed.

Rhododendrons and azaleas 2

Lichen growth

Leaf spot

Azalea gall

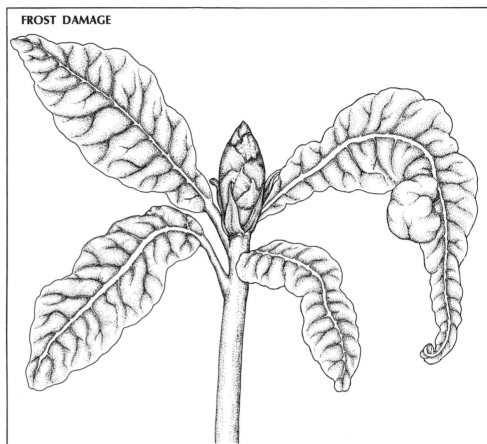

FROST DAMAGE

checking the pH of the soil before planting rhododendrons, and if it is very alkaline do not grow them.

Magnesium and manganese deficiencies may cause yellowing between the veins on plants growing in acid soils. When magnesium is deficient, the older leaves often show orange or purple blotches between the veins. When yellowing occurs check the pH of the soil and, if it is below 5.0, spray affected shrubs with a solution of $\frac{1}{2}$ lb of magnesium sulfate in 3 gal of water plus a spreader such as soft soap or a few drops of mild liquid detergent. Repeat once or twice at two week intervals, but do not spray when the plants are in flower. If the yellowing occurs on the younger leaves, then the plant may be deficient in manganese. In this case spray two or three times at two week intervals with a solution of 2 oz of manganese sulfate in 3 gal of water plus a spreader.

Leaves spotted

Leaf spot is a fungal disease that may be caused by any of a number of different genera and species. In all cases, they cause purple, tan or brown circular blotches to develop on the leaves of affected bushes. In severe cases cut off affected leaves and spray with benomyl, maneb or zineb. This trouble is worse on weak bushes, so spray with a foliar feed during the growing season and correct any adverse soil conditions such as drought, waterlogging or malnutrition by suitable treatment (see pages 5–7).

Leaves distorted

Azalea gall (*Exobasidium vaccinii*) affects evergreen azaleas and also azaleas grown as indoor pot plants. The young developing leaves and flower buds become pale green, or less frequently red, then swell up to produce small galls. These turn white and waxy as a

Frost may cause considerable damage to the leaves, stems, buds and flowers of rhododendrons, and tender varieties should not be planted in frost pockets. The leaves are only damaged when they are in bud or still small. As they develop they become distorted, curled and often grow only on one side of the main vein. They may also develop rows of small holes either side of the main vein. Affected leaves become increasingly distorted as they grow, although they usually become less noticeable later on as uninjured leaves develop.

Frost may cause the bark of affected stems to split longitudinally near ground level or, less frequently, on the young shoots, and in severe cases it peels away. These symptoms are often overlooked, and the damage will only be noticed some months later when affected shoots die back. If the shoots have been injured fairly recently, bind them up with grafting tape, remembering to cut it off once the tissues have healed. Later, apply a protective wound paint to any small splits on ungirdled shoots to prevent the entry of pathogens. If the shoots have died back cut them off 1 in below the split.

Flower buds affected by frost turn brown and become soft and rotten. If such a bud is opened, its petals may show some color but the stamens within the flower will be black. Remove affected buds.

81

Rhododendrons and azaleas 3

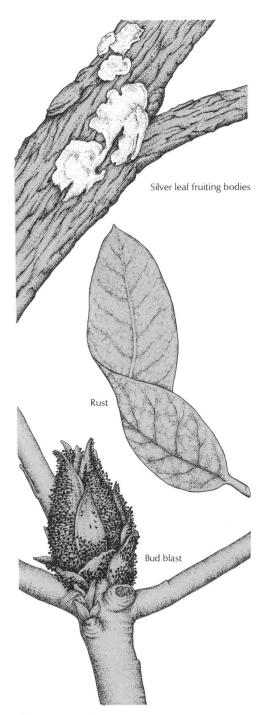

Silver leaf fruiting bodies

Rust

Bud blast

coating of fungus spores develops on them. Eventually they turn brown and shrivel. Pick off and burn the galls, if possible before they turn white. Then spray diseased plants with a copper fungicide or zineb and spray again the following season as the plants come into growth, repeating once or twice at two week intervals.

Leaves with obvious growths
Green algae, especially *Pleurococcus*, sometimes form unsightly thick green deposits on the surface of rhododendron leaves. Such growths develop on shrubs growing in very humid or sheltered positions. Control algae by spraying at monthly intervals with maneb or mancozeb. On small shrubs wash off the algae with soapy water and rinse with clear water. Where the affected shrubs are overgrown or surrounded by dense trees open up the area and prune the bushes to provide a good circulation of air around them to prevent the further development of green algae.
Rust (*Chrysomyxa rhododendri*) shows as pustules of orange and brown spores scattered on the undersides of leaves, the upper surfaces developing yellow patches. These symptoms are similar to those produced by rhododendron lacebug (see page 80). The fungus lives within affected leaves and, if left untreated reappears every season. Cut off and burn affected leaves, and spray plants with mancozeb.

Stems with visible growths
Lichens and mosses of many types can grow on rhododendrons. Lichens either appear as thin flat crusts or they resemble leafy or bushy plants growing on the bark. Mosses may form large, loose tufts, or may be densely compact cushions. Such growths sometimes develop on vigorous plants if the humidity is high, but elsewhere they are more often found on plants lacking in vigor, particularly those that are already dying back from some other cause. As there is no safe spray that will control these growths encourage vigor by spraying during the growing season with a foliar feed, by feeding in the spring and by mulching, watering and carrying out any other necessary improvements to the soil.

Prune out badly affected shoots once the plant has started to grow away well.

Stems dying back
Silver leaf (*Stereum purpureum*) can cause branches to die back or the complete death of plants, but, despite its name, no silvering of the foliage occurs. When cutting out dead branches, examine the cut if it is longer than $\frac{1}{2}$ in in diameter to see if there is a brown or purple stain in the internal tissues, since this indicates that the die-back has been caused by the silver leaf fungus. The stain can be seen more easily if the cut surface is moistened. Cut back all affected branches to about 6 in beyond the point where the stain ceases. Paint the wounds with a fungicidal paint. Feed, mulch and water the affected shrub to encourage vigor. If fruiting bodies of the fungus appear at the crown, destroy the affected plant. The fruiting bodies are purple when fresh, later becoming white or brown.
Honey fungus (*Armillaria mellea*) frequently kills rhododendrons. The first noticeable symptom is discoloration and drooping of the leaves which do not fall. The affected plant is usually dead within a week of such symptoms appearing, even though the shoots may still be green. The fungus shows as white fan-shaped sheets of fungal growth beneath the bark of the roots and around the crown of the shrub at or just above soil level. When fresh it has a smell resembling mushrooms, which distinguishes it from the gray-white deposit that often grows naturally beneath rhododendron bark. Dark brown root-like structures called rhizomorphs may be found on diseased roots; these grow out through the soil to spread the disease. Honey-colored toadstools may appear in the fall at the base of a dying shrub. Dig up and burn dead and dying shrubs together with as many roots as possible. Sterilize the soil with a 2 per cent solution of formalin applied at the rate of 6 gal per square yard. Try to trace the source of the infection and destroy any old stumps and woody debris in the soil.
Soil-borne fungi, particularly the species *Phytophthora cinnamoni*, can cause die-back of shoots. Affected shrubs usually show a brown stain beneath the bark at and just above ground level. Destroy affected plants

and sterilize the soil with a 2 per cent formalin solution at the rate of 6 gal per square yard. In container-grown stock, spraying with ethazol at monthly intervals will prevent the wilt disease.

Flower buds dying
Bud blast is caused by the fungus *Briosia azaleae*. The buds turn brown, black or silver in spring and, later, black bristle-like heads of fungus spores protrude from them. The buds do not rot but remain firm and may stay on the bush in this condition for two or three years. The best method of control is to cut them off (although they are not easily removed) and burn them if the bush is not too large and not many buds are affected. Help to prevent this disease by taking the appropriate measures to control rhododendron leaf-hoppers (see page 80) because infection is believed to occur through wounds caused by this pest when it lays its eggs in the bud scales.

Flowers dying
Petal blight of azaleas is caused by the fungus *Ovulinia azaleae*. Small spots develop on affected petals—these are white on colored flowers and pale brown on white blooms. As the spots increase in size they take on a water-soaked appearance and eventually the flowers are reduced to wet slimy masses. Only three days may elapse between the onset of the disease and the complete collapse of the flowers, so that the flowering period is severely curtailed. Affected blooms do not fall but hang on the bushes as trusses of withered flowers which look as if they have been attacked by frost. Small black seed-like objects called resting bodies appear in large numbers on the dead flowers, which can remain on the bushes in this condition until the following spring. Once the disease has appeared nothing can be done that season to control an affected shrub, but remove as many of the diseased flowers as possible to prevent the trouble spreading to later-flowering azaleas. Spray with benomyl as soon as the flower buds begin to show color and repeat at five day intervals throughout the flowering period. This should be routine treatment where this disease is known to be troublesome.

Broad-leaved woody plants 1

This section covers broad-leaved trees, shrubs and woody climbers, with the exception of roses, rhododendrons and azaleas, which have their own sections.

Leaves with pests visible

Aphids (many species) suck sap from the foliage and shoots of most trees and shrubs during the spring and summer months. This can cause the leaves to curl at the shoot tips, especially on flowering cherries and viburnums. Control aphids by spraying with dimethoate, formothion, menazon or pirimicarb as soon as they are seen; however, this is only worth while on those trees and shrubs that are small enough to be sprayed thoroughly. Many of the aphids that occur on woody plants over-winter as eggs, which are laid near the buds and in cracks in the bark. Destroy these eggs on deciduous trees and shrubs by applying a dormant oil in December or January.

Soft scales (*Coccus hesperidum*) are flat, oval, yellow-brown creatures about $\frac{1}{6}$ in long, and can be found on the undersides of leaves of bay, camellia, ivy and many other shrubs. They feed by sucking sap, and can be found clustered along the larger leaf veins. Control them by spraying the undersides of the leaves with malathion on two or three occasions at intervals of about 14 days between April and October.

Whitefly (various species) attack the leaves of honeysuckle, *Viburnum tinus* and certain other shrubs. The adults are small, white, moth-like insects, and are not the same whitefly species as those that attack brassicas and greenhouse plants. The larvae and pupae are flat, oval, scale-like objects which, like the adults, feed by sucking sap from the undersides of the leaves. Control whitefly by spraying thoroughly with a pyrethroid compound or pirimiphos-methyl when the adults are seen. Several applications at weekly intervals may be required.

Leaves eaten

Caterpillars of many species of moth and sawfly will damage the foliage of most trees and shrubs. The larvae of some species feed together in groups, and may completely defoliate small shrubs, but the effect on a woody plant is usually not too serious and can be tolerated. If controls are necessary, spray the affected tree or shrub with fenitrothion, trichlorphon or pirimiphos-methyl. Those species that have gregarious larvae can

sometimes be controlled by cutting out the affected branches. The small green caterpillars of tortrix moths feed out of sight at the shoot tips between leaves that the larvae have tied together with silken webbing. On small shrubs control these caterpillars by squeezing any webbed shoot tips. Extensive infestations may require spraying with pirimiphos-methyl or trichlorphon.

Leaves with holes

Shothole is caused by the fungus *Coccomyces hiemalis*, and attacks the leaves of flowering cherry and other *Prunus* species that are lacking in vigor. Numerous rounded brown spots develop on the leaves, and the dead tissues later fall away, leaving holes. Prevent this disease by feeding, mulching and watering trees to increase their vigor. Should symptoms appear, spray with a foliar feed during the growing season. If the disease recurs spray with benomyl, captan or dodine at petal fall and at ten day intervals.

Leaves distorted

Capsid bugs (*Lygus rugulipennis* and *Lygocoris pabulinus*) are active between late spring and the end of summer, with most damage occurring from May to July. Capsids are green or brown insects about $\frac{1}{4}$ in long which suck sap from the buds and young leaves, most frequently of forsythia, caryopteris, hardy fuchsia and hydrangea. They have a toxic saliva that kills plant cells in the immediate area of their feeding. This causes leaves, when fully expanded, to have many small holes and they may also be considerably distorted. Attacks on fuchsias often prevent the flowers from forming. Where trouble has been experienced in previous years, spray susceptible plants with dimethoate, formothion or fenitrothion or in late May, or earlier if signs of damage are noticed. Hydrangeas can be harmed by many insecticides, but malathion is safe if applied in the evening when temperatures are lower, and after the plant has been watered as necessary.

Frost damage causes the leaves of camellias, magnolias, fatsia and, occasionally, other shrubs to become distorted, curled and often one-sided as they develop. These symptoms may also be caused by faulty root action

Soft scale on bay

Frost damage on camellia

Buff tip moth caterpillar on hornbeam

Whitefly pupae on *Viburnum tinus*

Capsid bug damage on hydrangea

Broad-leaved woody plants 2

producing irregular growth when leaves are in bud or still small. Affected leaves become more and more distorted as they grow, although they are usually less noticeable later on when uninjured leaves develop. There is no treatment that can be carried out to improve their condition.

Leaves mined

Leaf miners (many species) are the larvae of flies, moths, sawflies and beetles. Examples include the larvae of the holly leaf-mining fly (*Phytomyza ilicis*) and the lilac leaf-mining moth (*Caloptilia syringella*), which also attacks privet. The former causes roughly circular blotch mines with a diameter of up to $\frac{1}{2}$ in. These are yellow or purple in color and are readily seen on the upper surface of the leaves, especially on clipped hedges where nearly every leaf may be affected. In spite of this, little harm seems to be done to the holly's vigor. Lilac leaf miner has two stages in its larval development. The young larvae live inside the leaves, causing blotch mines which turn brown and shrivel up. Later, when the caterpillars become larger, they emerge from the leaf and proceed to roll it up from the tip end, using silken threads to hold it in a rolled position. They complete their development by feeding on the rolled leaf. Light infestations of leaf miners can be tolerated or dealt with by picking off the affected leaves. Spray extensive attacks with pirimiphos-methyl or trichlorphon as soon as leaf mining begins.

Leaves galled

Gall mites (*Eriophyes*) affect many deciduous trees, although each mite species is usually specific to one particular type of tree. Two common gall mites are nail gall mite (*Eriophyes gallarumtiliae*), which causes red, finger-like projections from the upper surface of linden leaves, and acer pimple gall mite (*Eriophyes acericola*), which causes small red pimples on the upper surface of maple leaves. These galls may occur in hundreds on the leaves but no real harm seems to be caused to the trees. Other gall mites can cause symptoms such as rolling of the leaf margins or the excessive development of hairs on the leaf surfaces. Again no real harm is done.

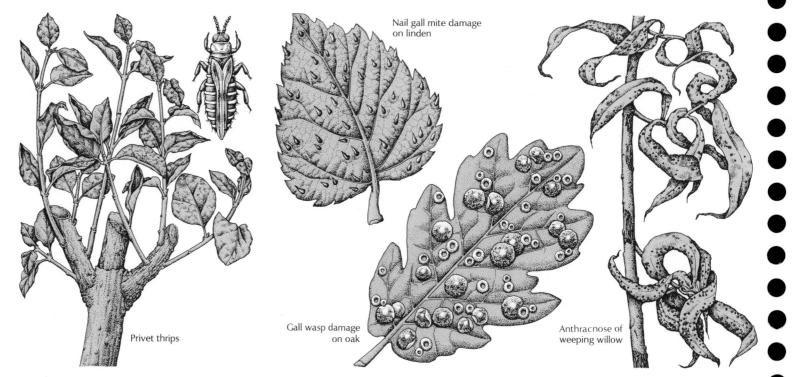

Privet thrips

Nail gall mite damage on linden

Gall wasp damage on oak

Anthracnose of weeping willow

Control gall mites by spraying with malathion when the buds open, repeating treatment two weeks later.

Gall wasps (Cynipids) are mainly associated with oak trees, where they form a variety of galls on the leaves, stems, acorns and roots. Towards the end of summer, several types of spangle gall may develop on the undersides of the leaves; these galls are rust-brown or brown disks, up to $\frac{1}{5}$ in in diameter. Although in some years galls may be extremely numerous, no harm is done to the tree, and control measures are not required.

Peach leaf curl (*Taphrina deformans*) affects only flowering peaches and almonds. All or part of the leaf becomes considerably thickened and curled, and blisters develop. These are red at first and later become covered with a pale bloom consisting of masses of fungus spores. On small trees, or where only a few leaves are swollen, reduce the infection by removing and burning these leaves before they turn white. Where a number of leaves are infected they fall very early in the season

and weaken the tree. Prevent infection by spraying the tree with bordeaux mixture or other copper fungicide in January or February. Repeat 10–14 days later and again just before leaf-fall. The disease cannot be cured by spraying when symptoms are seen.

Yellow leaf blister (*Taphrina populina*) affects only poplars, causing them to develop large yellow blisters on the lower leaf surface. Control measures are not usually necessary. However, if this disease is troublesome every year, and the tree is small enough to spray, apply either ferbam or zineb as a dormant spray in early spring. Ensure that the tree is thoroughly covered.

Leaves discolored

Lime-induced chlorosis occurs when the soil is so alkaline that plants are prevented from obtaining iron and manganese. Affected plants show yellowing between the veins and, in severe cases, the leaves at the shoot tips may be almost white. Most susceptible to this trouble are, among woody plants, camellias

and most heathers, and also all other plants that require acid soil conditions. Where the pH is over 7.5, symptoms may also appear on hydrangeas, ceanothus, chaenomeles and wisteria. Try to reduce the pH of the soil by digging in acidic materials such as peat, crushed bracken or pulverized bark, or apply acidifying chemicals such as sulfur or aluminum sulfate (see pages 5–7). Only use the latter chemical before planting. Rake in fritted trace elements or apply a chelated compound annually, and mulch each year with acidic materials. Do not grow acid-loving plants in very alkaline soils.

Magnesium and manganese deficiencies can cause yellowing between the leaf veins on plants growing in acid soils. When magnesium alone is deficient, older leaves often show orange, purple or brown blotches between the veins. Lilacs, *Prunus* spp and wisteria are most susceptible to these troubles. When yellowing occurs, check the pH of the soil and, if it is below 5.0 spray affected shrubs with a solution of $\frac{1}{2}$ lb of magnesium

Broad-leaved woody plants 3

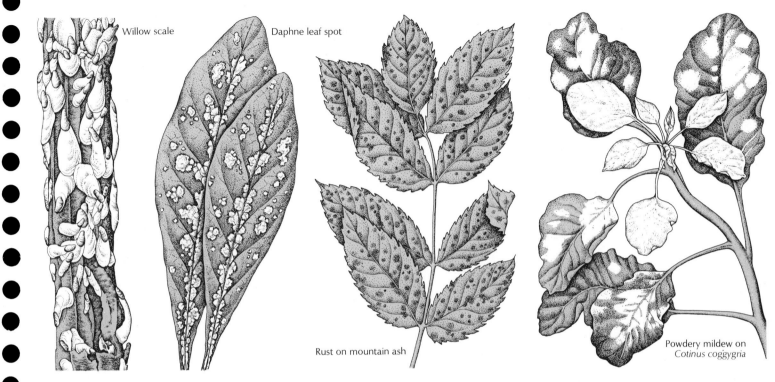

Willow scale

Daphne leaf spot

Rust on mountain ash

Powdery mildew on *Cotinus coggygria*

Privet thrips (*Dendrothrips ornatus*) commonly damage privet hedges and lilac bushes during late June to August. Thrips are thin, black or yellow insects about $\frac{1}{10}$ in long which suck sap from the upper surface of the leaves. This causes the foliage to become dull green. Damage is more likely to occur in warm dry summers, and can be prevented by spraying with dimethoate, formothion, pirimiphos-methyl or fenitrothion when the insects are seen.

Leaves spotted

Scab on crab apples is caused by *Venturia inaequalis*, and on pyracanthas by *Fusicladum pyracanthae*. Olive-brown blotches develop on the leaves and fruit, and may cause premature defoliation. Rake up and burn affected leaves. Spray crab apples at the green bud stage and after petal fall, and pyracanthas three times in March to April and twice in June to July, using captan, dodine, benomyl or thiophanate-methyl. Note that too frequent use of the last two fungicides could lead to the development of resistant strains of the fungus, so that the fungicides cease to be effective.

Leaf spot can be caused by many different fungi. The spots may appear as small brown blotches scattered over the leaves, or larger brown spots on which may show pinpoint-sized black fruiting bodies of the fungus. In severe cases the spots may coalesce until the leaves die, as with yucca, or premature defoliation may occur, as with daphne and poplar. Remove and burn affected leaves. Spray plants with zineb, maneb or mancozeb in the spring as the leaves open and repeat the treatment twice at ten day intervals. Prevent infection by planting carefully in well prepared soil and feed, mulch and water as necessary to maintain vigor. Applications of a foliar feed during the summer may also help to prevent infection or restore vigor to severely affected plants.

Anthracnose of weeping willows, especially *Salix babylonica*, is caused by the fungus *Marssonina salicicola*, and on sycamore by *Gnomonia platani*. Both diseases are most troublesome in wet springs and on sycamore can be confused with frost damage since developing leaves and shoots turn brown and

sulfate in 3 gal of water, plus a spreader. Repeat once or twice at two week intervals, but do not spray when the plants are in flower. If the yellowing occurs on the younger leaves, then the plant may be deficient in manganese. In this case spray two or three times at two week intervals with a solution of 2 oz of manganese sulfate in 3 gal of water, plus a spreader.

Frost damage or cold winds can cause the leaves of skimmia and magnolia to turn yellow, and those of garrya, camellia and bay to turn brown or bronze. Plants lacking in vigor from malnutrition or the soil being too wet or too dry are more susceptible to this type of injury, so try to prevent it by feeding, mulching, watering and draining as necessary. If damage has already occurred, applications of a foliar feed during the growing season may hasten the plant's recovery, but leaves that have already been affected will not recover.

Scorch is usually caused by cold winds or frost damaging soft tissues, and occurs most

frequently on acers and beech in spring. The leaves turn brown, especially around the margins, and then curl and shrivel to give a scorched appearance. Plants lacking in vigor due to poor planting, malnutrition or an irregular supply of moisture in the soil are most susceptible to this trouble. Prevent it by not planting susceptible trees and shrubs in a frost pocket. If necessary drain the soil before planting, plant carefully and feed, mulch and water to maintain vigor. If damage has already occurred, apply a foliar feed during the growing season. As with frost damage, affected leaves will not recover.

Faulty root action due to poor planting, malnutrition or the soil being too wet or too dry can affect all types of woody plants. Leaves and stems of aucuba turn black, and those of other plants turn yellow or brown. Affected leaves often fall prematurely. Whenever leaves are discolored, consult the section on physiological disorders, pages 5–7.

Silver leaf (*Stereum purpureum*) can affect many types of woody plants but is most

common on species and hybrids of *Populus, Prunus, Lonicera, Azara* and *Cotoneaster*. The fungus enters through wounds such as pruning snags, and causes the leaves on one or more branches to become silvered and then sometimes brown. Affected branches die back progressively and small fruiting bodies of the fungus develop on the dead wood. These are purple when fresh, later becoming white or brown. They may be bracket-shaped, overlapping, or they may lie flat. A brown or purple stain develops in the internal tissues and is easily seen if the cut end of an affected branch at least an inch in diameter is moistened. Cut back all dead branches to about 6 in beyond where the stain ceases, preferably during summer when there is the least chance of new infections occurring. Sterilize tools before and after use, and paint the wound with a fungicidal paint. Destroy the tree if fruiting bodies of the fungus appear on the trunk. Otherwise feed, mulch and water the infected tree to encourage vigor.

Broad-leaved woody plants 4

die. Later, characteristic brown banding appears along the veins; the rest of the leaf at first remains green but eventually the entire leaf is killed and falls prematurely. Willow anthracnose shows as small brown spots on the leaves, which are often distorted and fall prematurely in vast numbers. Both willows and sycamores develop small cankers on the shoots and in severe cases the shoots die back. Rake up and burn fallen leaves. Spray affected trees with a copper fungicide, dodine, maneb or chlorothalonil as the leaves unfold in the spring, and repeat twice during the summer. Where large trees of great value are affected it may be worth while employing a tree surgeon who will be able to treat the tree with chemicals that are not available on the amateur market.

Leaves with corky patches
Oedema, or dropsy, on outdoor plants is usually caused by the soil being too wet. It shows as small corky outgrowths on the leaves of camellias and eucalyptus. Do not remove affected leaves since this will only make the trouble worse. No special treatment can be given but with correct cultural treatment (see pages 5–7) affected plants should recover in due course. Affected leaves will, however, remain in this condition for the rest of their lives.

Leaves with visible fungal growth
Powdery mildews can be caused by many species of fungus. In all cases a floury white coating of fungus develops on the leaves. Young shoots can be severely injured and should be cut out in the fall. At the first signs of trouble spray with benomyl, dinocap, thiophanate-methyl or sulfur, repeating as necessary. As plants are more susceptible to infection when they are dry at the roots, try to prevent infection by mulching and by watering in dry periods before the soil dries out completely.
Downy mildew (*Peronospora grisea*) affects *Hebe* species. This mildew causes the upper surfaces of leaves to develop pale blotches, and grayish fungal growths appear on the lower surfaces. Remove the most severely affected shoots and spray with mancozeb or zineb.
Rust is caused by various fungi and attacks box, birch, mahonia, hawthorn and species of *Sorbus* such as mountain ash. Spore pustules develop on the undersides of the leaves, which show discolored patches on the upper surfaces. Occasionally, the pustules may also appear on the upper surface. The spore masses are usually orange at first and may become brown later. On hawthorn the pustules appear as swellings and may also affect the shoots and fruits. Severely affected leaves on any host may fall prematurely and should be raked up and burned, otherwise the rust fungus will over-winter on them. Try to prevent infection by feeding and mulching, and by watering in dry periods; these are the only measures that can be used for tall trees. On small shrubs, remove affected leaves and spray with maneb, mancozeb or zineb. Do not grow *Sorbus* species near *Juniperus* species as both hosts can be severely attacked by the same rust (see also page 92 for details of rust on *Juniperus*).

VIRUS DISEASES OF TREES AND SHRUBS

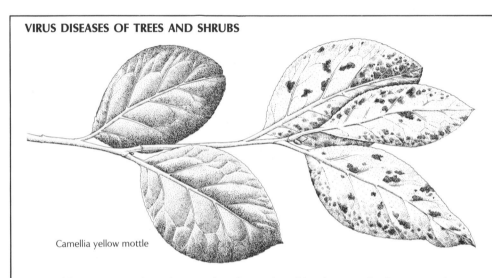

Camellia yellow mottle

Many different viruses have been isolated from unhealthy trees and shrubs, the most common being arabis mosaic virus and cucumber mosaic virus. In general, viruses cause yellow mottling and blotching of leaves, which may become distorted and develop line or ring patterns. Banding of the veins may also occur. The most severely affected plants are buddleia, which shows a narrowing of the leaf blade, and *Daphne odora* and *D. mezereum*, which can show almost complete yellowing of the leaves, with plants becoming stunted and failing to flower. Clematis and passion flower can also show severe symptoms, with plants becoming stunted, bearing distorted leaves and flowers, and possibly ceasing to flower a year or so after being infected.

Some viruses are only seen to affect one type of plant; thus hydrangea ringspot virus has only been found in hydrangeas, where it causes crinkling, rolling or distortion of the leaves, with fewer and smaller florets appearing on short flower stalks. Hydrangeas can also be affected by a viral-like organism known as a mycoplasma that causes flowers to remain green. Camellias may be affected by a viral-like disease known as camellia yellow mottle. Diseased plants show yellow or white blotches on the leaves, and some leaves may be completely white.

Whether plants suffering from virus attacks show discolored leaves depends to a certain extent on the weather conditions: in some seasons affected plants show no symptoms and appear healthy. It is not necessary, therefore, to destroy all infected woody plants; only those that are severely affected should be destroyed, particularly if the disease attacks the flowers. However, if an infected plant is retained it must be remembered that the virus could spread to other plants in the garden and cause more severe symptoms on them.

After digging up an infected tree or shrub, burn it and sterilize the soil with a 2 per cent solution of formalin applied at 6 gal per square yard to kill eelworm carriers. When replanting, do so with a plant of a different type and do not plant daphne, forsythia, ivy or spiraea close to a privet hedge since it may harbor arabis mosaic virus, to which they are susceptible.

Do not propagate from any tree or shrub showing signs of virus infection. After handling or cutting a diseased plant wash hands and tools with soapy water before touching a healthy plant.

Broad-leaved woody plants 5

Stems with pests visible

Scale insects of various types occur on the bark of many trees and shrubs. Some, such as oystershell scale (*Lepidosaphes ulmi*) and willow scale (*Chionaspis salicis*), are tiny flat pear-shaped objects about $\frac{1}{10}$ in long which form dense colonies on the older branches. Others, such as brown scale (*Parthenolecanium corni*) are oval, convex, red-brown objects up to $\frac{1}{4}$ in long. Heavy infestations of these sap-feeding insects will reduce the plant's vigor and may contribute to dieback of branches. On deciduous trees and shrubs the easiest means of control is to apply a dormant oil in December or January. Alternatively, spray the plants twice with malathion at 14 day intervals when scale insect nymphs are hatching out. This normally occurs sometime between May and September, and varies according to which species of scale insect is involved.

Cuckoo spit is caused by the nymphs of froghoppers such as *Philaenus spumarius*. These creamy-white insects occur in May to June on the soft stems of many shrubs, and are hidden underneath a white, frothy liquid known as cuckoo spit, which these sap-feeding insects secrete. Little harm is caused by either the nymphs or the adults, although the nymphs may cause some leaf distortion if they are feeding at the shoot tip. Control measures are not normally necessary, but if cuckoo spit is considered unsightly spray the plants with dimethoate or formothion.

Stems with visible growths

Lichens and mosses of many types can grow on trees and shrubs. Lichens appear either as thin flat crusts or they resemble leafy or bushy plants growing on the bark. Mosses may form large loose tufts, or may be densely compact cushions. Such growths sometimes develop on vigorous plants if the humidity is high, but otherwise they are more often found on plants lacking in vigor, particularly those that are already dying back from some other cause. There is no safe spray that will control these growths on evergreen trees and shrubs. Spray deciduous plants during their dormant period in winter with a 5 per cent dormant oil wash. Try to encourage vigor in affected trees and shrubs by feeding

in the spring, and by mulching, watering and carrying out any other necessary improvements to the soil. Once an affected evergreen has started to grow away well, prune out badly affected shoots.

Stems splitting

Frost damage may cause longitudinal splitting of the bark on young shoots of trees and shrubs if they grow in frost pockets, camellias being most susceptible to this trouble. On other plants the splitting usually occurs towards ground level and, in severe cases, the bark peels away in strips from the underlying wood tissues and complete girdling may occur. Unfortunately, these symptoms are often overlooked and the damage may not be noticed until the affected shoots die back some months later. When the shoots have been injured by frost fairly recently bind them up with grafting tape and cut it off after the tissues have healed. Later, apply a protective wound paint to any small splits where girdling has not yet occurred since they may then start to heal over naturally. Where die-back has occurred, cut out affected shoots an inch or so below the split; the shrub may then shoot again from the base.

Stems galled

Broom gall mites (*Aceria genistae*) are microscopic animals that live inside buds of cytisus, causing them to form cauliflower-like galls. The galls are initially green and soft, but by the end of summer they become dry and brown. The galls seem to occur most commonly in the center of the bush, although all branches are attacked eventually and the plant's growth suffers. None of the chemicals currently available to amateur gardeners controls this type of mite. Light infestations can be held in check by picking off and burning the galls. Burn badly galled plants and, if replacing with new bushes of cytisus within 12 months, plant them in a different part of the garden to prevent reinfestation.

Crown gall (*Agrobacterium tumefaciens*) can affect daphnes and viburnums, causing a chain of small galls to appear on the shoots. Cut off affected shoots and burn them. Destroy severely affected plants. When plant-

Forsythia gall

Crown gall on daphne

ing another tree or shrub in infected soil, first dip the roots in a copper fungicide, and take care not to injure them.

Forsythia gall (cause unknown) is a disorder of forsythias, especially *Forsythia* × *intermedia*, that shows as nodular outgrowths on the stems. Cut off affected shoots and destroy them, since otherwise the galls may spread. The cause of forsythia gall is not known, but in some cases it is thought to be a form of bacterial crown gall.

Stems distorted

Fasciation is caused by an early injury to the growing point following insect attack or, more commonly, frost damage. Affected shoots become flat, although they still bear leaves and flowers. This trouble occurs most commonly on *Forsythia* and *Prunus subhirtella* 'Autumnalis'. Cut off affected shoots an inch or so behind the point where the flattening commences.

Witches' brooms can be caused by certain rusts, species of the fungus *Taphrina* or other organisms such as mites. Very occasionally they occur as a result of a mutation in the plant tissues. Each "broom" is caused by many shoots all growing abnormally from one point, usually in an erect manner, making them easily visible among the larger branches.

Cut out the affected branch to a point at least 6 in below the broom. No chemical control is effective.

Stems with peeling bark

Papery bark is caused by faulty root action, particularly waterlogging, and is most common on viburnums. The bark becomes paper thin and peels off. The shoots may also be affected and, if they have been girdled, dieback frequently occurs. Cut out all dead shoots, and on larger branches remove dead and rotting tissues beneath the peeling bark. Cover all wounds with a fungicidal tree paint and improve the cultural conditions to prevent further trouble.

Stems with slimy patches

Slime fluxes are caused by various unidentified fungi, yeasts and bacteria. On large trees, particularly beeches, birches and oaks, slime fluxes usually show as black oozing patches of dead bark on the trunks. In the case of clematis, especially *C. montana*, affected shoots become covered with a cream, pink or orange slimy growth. The original cause of most slime fluxes is a wound through which some of the plant sap escapes, especially if there is excessive water in the trunk of the tree accompanied by high gas

Broad-leaved woody plants 6

pressures, a trouble known as "wetwood". Occasionally, organisms ferment this sap, which injures the surrounding bark tissues and makes the trouble progressively worse. Where a slime flux occurs on a large tree employ a tree surgeon to deal with it. On a small tree remove all discolored and decaying bark and inner tissues to leave a clean wound. If sap continues to ooze out of the initial wound, cauterize the tissues with a blow lamp. Then paint the whole area with a wound paint. If the original wound was a crack in the bark caused by irregular growth due to drought followed by waterlogging, good cultural treatment and the use of a foliar feed on small trees should aid recovery. On clematis, cut out affected shoots to an inch or so below the diseased tissues; with clematis plants the trouble usually follows frost damage. Nevertheless, they should be sprayed with a copper fungicide in case clematis wilt (see page 89) was the original cause of the slime flux.

Stems cankered

Canker affects crab apples, ash, beech and mountain ash, and is usually caused by *Nectria galligena*. (Note that beeches can also be affected by *Nectria coccinea*, which causes beech bark disease. This trouble is not common in gardens, however.) Canker shows as sunken and discolored patches on the bark. As they extend they become elliptical, with the bark shrinking in concentric rings around each canker. The branch usually becomes swollen around the canker and girdling of the shoot may occur, causing dieback. In summer, white pustules of fungus spores form on the sunken bark and, later, small red fruiting bodies develop by which the fungus over-winters. Infection can occur at any time of the year through wounds, pruning cuts and leaf scars. Cut out severely cankered branches and spurs. On large branches where no die-back has occurred cut out the brown diseased tissues with a chisel or sharp knife, burning the parings. Paint the wounds with a thiophanate-methyl paint. Since the disease is worse on trees that are lacking in vigor, feed and mulch them and improve the drainage if much waterlogging occurs.

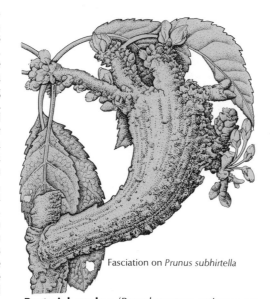

Fasciation on *Prunus subhirtella*

Bacterial canker (*Pseudomonas syringae* or *P. mors-prunorum*) affects flowering almonds and cherries. The cankers show as elongated flattened lesions from which exude copious amounts of gum. The following spring the buds of an affected branch usually fail to open. If leaves do develop, they turn yellow and become narrow and curled, then wither and die during the summer as the branch dies back. Another symptom of the disease is brown circular spots on the leaves with the affected tissues falling away to leave holes resembling those of shothole (see page 83). Very occasionally the bacteria die out in the cankers and no further symptoms are seen. Nevertheless, remove badly cankered branches and dead wood, and paint the wounds with a fungicidal wound paint. The bacteria live on the leaves during the summer, therefore, spray affected trees with a copper fungicide in mid-August, mid-September and mid-October.

Bacterial canker of poplars (*Aplanobacter populi*) occurs on many poplar species, though not on *Populus nigra*. Unsightly cankers up to 6 in in length develop on shoots, branches and sometimes the trunk. Young shoots may die back in early summer and, on one-year-old shoots, bacteria ooze out of cracks in the wood as a dull cream-

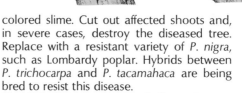

Canker on poplar

Papery bark

Anthracnose

colored slime. Cut out affected shoots and, in severe cases, destroy the diseased tree. Replace with a resistant variety of *P. nigra*, such as Lombardy poplar. Hybrids between *P. trichocarpa* and *P. tacamahaca* are being bred to resist this disease.

Fireblight (*Erwinia amylovora*) affects chaenomeles, cotoneaster, hawthorn, pyracanthas, sorbus and stranvaesia. The disease causes cankers to appear at the base of dead shoots in the fall, and if the bark is pared off a red-brown discoloration can be seen in the inner tissues. The bacteria usually enter through late spring or summer blossoms in warm weather, and spread down the spurs and lateral twigs into the main branches, causing die-back and browning of the leaves, which become withered but do not fall. In spring the cankers become active again and exude droplets containing bacteria. Control fireblight by pruning out all cankered branches back to a point at least 6 in below the apparently affected tissues, disinfecting all pruning tools after use. Then spray the affected plant with streptomycin when the flowers are at the ballon stage (ie shaped like a brandy glass). Repeat the treatment five days later. During extended flowering periods repeat this treatment again after another five days have elapsed.

Anthracnose is caused by *Marssonina salicicola* on weeping willows and *Gnomonia platani* on sycamore trees. Both cause small cankers to appear on shoots. In the later case these are oval and develop at infected nodes where buds fail to open in the spring. The fungus can grow back into the stem and, by girdling it, cause die-back of the shoot. In severe cases young developing shoots up to 4 in in length wilt and die, and the leaves turn yellow or brown and fall prematurely. In willow anthracnose, very many small dark cankers about $\frac{1}{4}$ in in length develop on the shoots, sometimes only an inch or so apart. Severe die-back often occurs and when many of the young growths are affected, the weeping habit may be lost. For control measures see under Leaves spotted, page 85.

Stems with small eruptions

Tan bark is a physiological disorder that shows on one-year-old shoots of flowering cherry trees, or sometimes the main trunk, as small eruptions containing a tan-colored powder. These eruptions are the breathing pores (lenticels) of the tree and are prominent naturally, but on affected trees they burst open causing the outer layers of the stem to peel off to expose brown rust-like masses of dead plant cells. Despite being unsightly,

Broad-leaved woody plants 7

Honey fungus

Lilac blight

this trouble is not serious since it usually occurs where the root action is vigorous, although it can also be caused by water-logged soil. Apart from draining the soil, no treatment is required unless much of the bark has peeled, in which case remove the loose bark and also any dead tissues beneath to leave a clean wound. Cover this with a fungicidal paint.

Stems dying back

Honey fungus (*Armillaria mellea*) is the commonest cause of die-back of trees and shrubs, particularly of privet hedges. Affected plants may die rapidly following the onset of the first obvious signs of trouble; usually this is either discoloration and withering of the leaves, which do not fall, or failure of the buds to open in the spring, although the shoots are still green and appear to be alive. The fungus shows as white fan-shaped growths appearing beneath the bark of the roots and trunk at, or just above, soil level. Dark brown root-like structures known as rhizomorphs are found on the diseased roots and grow out through the soil to spread the disease. Another symptom is the appearance in the fall of honey-colored toadstools at the base of a dying tree or shrub. Dig up and burn dead and dying trees and shrubs together with as

many of the roots as possible. Sterilize the soil with a 2 per cent solution of formalin applied at the rate of 6 gal per square yard.

Soil-borne fungi other than honey fungus, such as *Phytophthora* spp and *Verticillium* spp, can also cause die-back of shoots. The former is most troublesome in heathers and the latter on acers and *Cotinus coggygria*. Affected plants usually show a brown stain beneath the bark at, and just above, ground level. If only one or two shoots are dying back cut them out and drench the plant at ground level with benomyl. Destroy the plant if the whole of it is affected, as is usually the case with heathers affected by heather wilt. Sterilize the soil with formalin solution as recommended for honey fungus (see above).

Clematis wilt is caused most commonly by the fungus *Ascochyta clematidina*. One or more shoots wilt and die very rapidly, often right down to the base. Jackmanii types, and some other large-flowered garden varieties, are most susceptible to the trouble, but it can occur on species, for example, *Clematis montana*. Do not remove an affected plant since it is likely that new shoots will develop below the wilted region, or even below soil level, either later in the season or the following spring. Cut out all the wilted shoots back to clean living tissues, even if this means

going down below the soil surface. Paint the wounds (however small) with a good protective paint, and spray new developing shoots with a copper fungicide or spray with benomyl.

Lilac blight (*Pseudomonas syringae*) shows first as angular brown spots on the leaves. Soon afterwards the young shoots blacken and wither away, and the symptoms may then be confused with frost damage. The bacteria that cause this disease enter the leaves through their breathing pores and then spread down into the shoots where they form pockets of infection in the outer tissues and from which reinfection will occur in subsequent years. Cut out affected shoots back to a healthy bud and spray with a copper fungicide. The following spring spray affected plants again with a copper fungicide as the leaves unfold.

Dutch elm disease (*Ceratocystis ulmi*) is spread by the elm bark beetle. It causes the shoots to die, becoming hooked at the tips, and the leaves to turn yellow then brown, and hang in a withered condition. There is no control but resistant varieties are being bred and may become available to amateur gardeners in due course.

Frost damage can result in die-back of shoots, particularly of young soft growth of acers, magnolias, ceanothus, griselinia and camellias. On camellias affected stems usually split first (see Stems splitting, above). On other plants, affected shoots usually show blackened foliage and may also develop gray mold. Cut out affected shoots an inch or so below the dead tissues, and do not plant susceptible shrubs in a frost pocket.

Die-back can have a variety of causes, and affected plants should be examined for symptoms that indicate its origin. If no definite symptoms can be seen, it is probably caused by faulty root action from poor planting, malnutrition or unsuitable soil conditions. For treatment, see pages 5–7.

Stems with visible fungal growth

Powdery mildews show on both shoots and leaves as a white mealy coating of fungus spores. Affected shoots are often stunted and may die back at the tips; signs of fungal growth may still be seen in the winter. Cut out

affected parts to a point two or three buds below the apparently diseased tissues. For chemical control see under Leaves with visible fungal growth, page 86.

Coral spot (*Nectria cinnabarina*) is a very conspicuous disease in damp weather when it shows on the bark of dead shoots as numerous small pink or red, cushion-like pustules. Species of *Magnolia*, *Elaeagnus*, *Cercis* and *Acer* are particularly susceptible to infection, and may be killed completely if the fungus enters the main trunk or branches of the plant. Cut out all dead wood to 4–6 in below the apparently diseased tissues, and paint larger wounds with a protective paint. Burn all woody garden debris.

Gray mold (*Botrytis cinerea*) shows as gray velvety cushion-like pustules of spores on dead shoots, particularly of ceanothus, lavender, magnolia and lilac. The fungus usually enters through wounds caused by frost, or through tissues that are dying from faulty root action. Cut out affected shoots to at least 2 in below the apparently diseased tissues. Check whether the plant has been affected by any trouble at the roots and carry out any necessary remedial measures.

Bracket fungi (various species) attack all types of trees, especially ash, alder, beech, oak, poplar and birch (which is the most commonly attacked). The fruiting bodies of the fungi resemble shelves or brackets, and grow out of the side of trunks and branches, sometimes fairly high up and at other times appearing at ground level. They can be up to a foot or so across, and may not appear until months, or even years, after infection took place. Once they have shed their spores, the fruiting bodies may disintegrate fairly quickly, or they become perennial, remaining as hard lumps on the trunk. Remove any small branch bearing a bracket fungus, cutting back into clean wood. Cut out any brackets on the main trunk of a tree, and also any rotting wood beneath. Any old and large tree bearing fruiting bodies of a bracket fungus should be examined by a tree surgeon to determine whether it is safe, and can be treated, or should be felled. A birch tree showing the fruiting bodies should be felled since it will very likely be killed by the fungus.

Broad-leaved woody plants 8

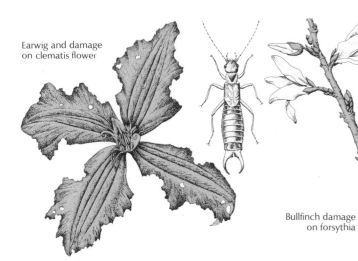

Earwig and damage on clematis flower

Bullfinch damage on forsythia

Pyracantha scab

Bracket fungi on birch

Flower buds dropping

Bud drop is usually due to the soil being too dry when the flower buds are just developing, particularly in late summer and early fall. It is prevalent on plants that are growing against walls; those most frequently affected are camellias, hibiscus and wisteria, which may also suffer from bud drop if it is cold at night. Mulch plants well to conserve moisture, and water in dry periods. Do not plant wisterias where they are likely to be affected by cold winds.

Lack of flower buds

Poor flowering or complete lack of blooms may be caused by bud drop (see above). If no flower buds form, incorrect pruning may be responsible, or growth may be too soft from over-feeding with nitrogenous fertilizers. Rake in a dressing of sulfate of potash at $\frac{1}{2}$ oz per square yard when the flower buds are developing. If the plant still fails to flower consider whether it is growing in too shady a position, or is suffering from faulty root action (see below). If necessary lift and re-plant carefully in a more suitable position. Trees and shrubs grown from seed may fail to flower because they are poor stock genetically; such plants should be destroyed.

Flower buds or flowers dying

Frost damage can cause browning of the flower buds, which become soft and rotten.

Camellias and magnolias are particularly susceptible. If an affected bud is opened the petals may show some color but the stamens within the flowers will be black. At a later stage of growth browning of the petals may occur. Remove affected buds and flowers, where possible, and do not plant tender trees and shrubs in a frost pocket. When severe weather is forecast protect small plants such as tree paeonies with old net curtains or sheets of newspaper.

Faulty root action is caused by drought, waterlogging, malnutrition or poor planting and can result in flower buds shriveling up and dying before they open, particularly on *Viburnum carlesii* and other species, and *Deutzia*. The buds do not usually fall, but remain on the bush in a withered condition. Feed, mulch and water as necessary, or drain the soil if much waterlogging occurs. Alternatively, lift the shrub and replant it more carefully. For details, see the section on physiological disorders, pages 5–7.

Blossom wilt (*Monilinia* spp) can affect ornamental *Prunus* spp and is a seasonal disease, being most troublesome in wet springs. All the flower tissues on an infected shoot wither, but do not fall. The leaves may also shrivel, and the tip of the shoot may die back. Cut out and burn all infected shoots in summer. The following season spray with benomyl as the first flowers open and repeat seven days later.

Paeony wilt (*Botrytis paeoniae*) can cause rotting of tree paeony buds, which become covered with a gray velvety growth of fungus. Cut off affected buds to an inch or so below the apparently diseased tissues. Spray with dichlofluanid soon after the leaves emerge; repeat at intervals of two weeks until the trees come into flower.

Honey fungus (*Armillaria mellea*) is often first noticed on flowering trees such as lilacs when the flowers do not open but remain as tight clusters of flower buds, which eventually wither. At this stage the shoots may still appear to be green and alive. For other symptoms and control measures see under Stems dying back, page 89.

Buds or flowers eaten

Bullfinches (*Pyrrhula pyrrhula*) feed on the flower buds of flowering cherry, forsythia, amelanchier and almond during the winter months. When frost or snow is present it is often possible to see discarded outer bud scales in large numbers beneath such trees, and an examination of the branches will show that many of the flower buds have been nipped off. Such losses become obvious at flowering time when the branches have relatively few blooms, which are usually situated at the outer tips of the branches. On small trees and shrubs the buds can be protected by netting, but for larger trees there is no effective protection. There are a number of

proprietary bird repellant sprays which may help but several applications are necessary during the winter, especially in wet years. Repellants may be successful while there are alternative sources of food available, but they do not work against very hungry birds.

Earwigs (*Forficula auricularia*) frequently destroy the flowers and young leaves of clematis. Since they feed at night their presence may go undetected unless the plant is examined by torchlight. Control them by spraying with either fenitrothion or trichlorphon at dusk on a warm still evening when damage is occurring. Several applications may be needed during the flowering period.

Fruits absent

Lack of pollination is a common cause of fruits being absent. It is due to adverse weather conditions discouraging pollinating insects and nothing can be done to prevent it. It can also be caused by the buds or flowers being damaged (see above).

Fruits blemished

Scab may attack crab apples and pyracanthas, the causal organisms being *Venturia inaequalis* and *Fusicladium pyracanthae* respectively. The fruits develop brown or black scabs, though on pyracanthas they may be reduced to clusters of small blackened fruits. Prevent infection by spraying as recommended under Leaves spotted, page 85.

Conifers 1

Leaves with pests visible

Woolly adelgids (*Adelges* spp) are sap-feeding insects that attack the foliage and stems of *Abies, Larix, Pseudotsuga* and *Pinus* species. They secrete white, waxy fibers over their bodies that completely hide them from view. Adelgids are not easy to control but, fortunately, conifers seem able to support large populations without any obvious ill effects. If the trees are small enough to be sprayed thoroughly, check these pests by spraying with malathion on a mild day in February or March, since at that time the over-wintering nymphs do not have a protective coating of waxy fibers. Repeat treatment in May to kill any remaining adelgids.

Leaves webbed

Juniper webber (*Dichomeris marginella*) is a small moth whose caterpillars live in colonies among the foliage of juniper trees. The caterpillars, which are chocolate-brown and up to $\frac{5}{8}$ in long, bind the leaves and branches together with silken webbing. They feed on the foliage and cause damaged leaves to turn brown, which gives the impression of dead patches on the bush. It is not easy to penetrate the webbing with insecticides but some control can be achieved by spraying forcefully with trichlorphon, pirimiphos-methyl or fenitrothion in May or June when caterpillars are seen.

Leaves discolored

Conifer red spider mites (*Oligonychus ununguis*) are tiny yellow-green animals that occur mainly on spruce (especially *Picea albertiana* 'Conica'), but may also attack juniper and thuja. Damage is most likely to be seen in hot dry summers when the leaves develop a mottled discoloration and defoliation may occur. Seen overall, the foliage appears yellow-brown. If the branches are examined with a hand lens, it is usually possible to see the mites and their spherical red eggs— these are laid among fine silken threads that cover the stems and foliage. Spray thoroughly with malathion, dimethoate or formothion on three occasions at seven day intervals as soon as the mites are detected.

Weather damage may occur following cold winds or frosts, the former causing browning

of *Sequoia* foliage, and the latter causing bronzing of *Thuja* foliage. No treatment can be carried out to prevent these troubles. Young small plants that are severely affected may respond favorably to applications of a foliar feed during the growing season, but affected plants usually grow out of the symptoms in the spring.

Faulty root action due to poor planting, malnutrition or the soil being too wet or too dry can cause browning of the foliage on most conifers. However, yews growing in waterlogged soil frequently show yellowing of the leaves, which usually fall prematurely. For remedial treatment, consult the section on physiological disorders, pages 5–7.

Needle cast may affect larch, pines, picea and pseudotsuga. Various fungi may be responsible and can cause considerable defoliation. On larches, affected needles turn brown in May and fall. Needles on the other conifers show a purple-brown discoloration and bear noticeable fruiting bodies of the fungus. These are black and spherical on picea and pines, and elongated orange-brown pustules in pseudotsuga. Rake up fallen needles and burn them since the fungi can survive on them for some time. Spray affected larches with chlorothalonil when the foliage is appearing in the spring, repeating at two or three week intervals until the end of July. Spray diseased pines with chlorothalonil at the end of July, August and September, or earlier if signs of infection are present. No chemical treatment is necessary for picea and pseudotsuga, but encourage vigor in affected trees by feeding, mulching and watering as necessary. Applications of a foliar feed during the growing season may also help to prevent further infection.

Rust fungi of various species can affect the needles of abies, larch and picea. Small pale blisters or bladder-like pustules containing yellow spores develop on the needles. These do not cause much harm, except that some defoliation may occur on picea, and control measures are not needed.

Stems splitting

Longitudinal cracking of the trunks of *Abies* is liable to occur in a hot dry summer. Prevent this by mulching to conserve mois-

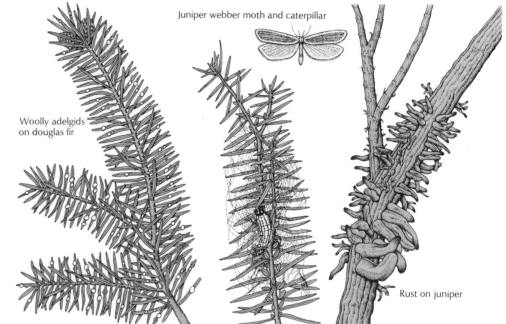

Juniper webber moth and caterpillar

Woolly adelgids on douglas fir

Rust on juniper

ture and by watering in dry periods. In most cases the wounds will heal naturally, but in severe cases clean the wounds by cutting out any rotting tissues, then cover with a wound paint.

Stems distorted

Witches' brooms may be caused by the rust fungus *Melampsorella caryophyllacearum* on *Abies*. On other conifers, however, they usually occur as a result of a mutation in the plant tissues. Each "broom" is caused by a number of shoots all developing abnormally from one point on the affected branch. The crowded mass of shoots is usually erect, and can be seen throughout the year among the larger branches. Cut out the branch bearing the broom to a point at least 6 in below the mass of shoots.

Stems cankered

Phomopsis disease is caused by the fungus *Phomopsis pseudotsugae* on pseudotsuga, the most badly affected trees, and by other species on cedar and larch, other conifers being only rarely affected. The fungi can live

on dead wood, and may also attack living trees through wounds such as those caused by frost. Once they have gained an entry they cause shoots to become girdled and die back. The main stem may be affected lower down and, in young trees, the part above the diseased area becomes swollen and the top of the tree dies. In older trees a canker develops, but the tree does not necessarily die. Small black fruiting bodies of the fungus are produced on cankers, dead twigs, dead patches of bark or living shoots. Cut out and burn all dead wood, and paint wounds with a protective paint.

Larch canker (*Dasyscypha willkommii*) is most common on *Larix decidua* and shows as large cankers on the branches or main trunk of the tree. Each canker is somewhat flattened and, if it girdles the branch, die-back occurs. Orange saucer-shaped bodies about $\frac{1}{8}$ in across appear on stalks at the edge of the cankers. Cut out dead shoots and cankered wood back to clean living tissues, and paint the wounds with a protective paint. Feed, mulch and water as necessary to encourage vigor.

Conifers 2

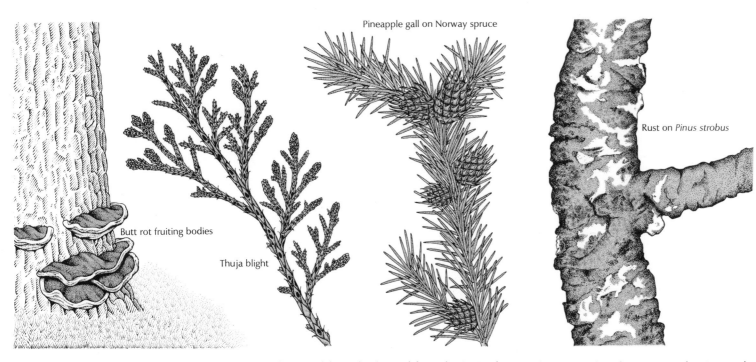

Pineapple gall on Norway spruce

Butt rot fruiting bodies

Thuja blight

Rust on *Pinus strobus*

Stem canker of *Picea* can be caused by several different fungi, but the commonest is *Nectria cucurbitula*, which produces red fruiting bodies on dead bark. The fungi enter through wounds, usually following frost damage, and girdle shoots causing die-back. Cut out dead shoots and paint all wounds with a protective paint.

Stems dying back
Honey fungus (*Armillaria mellea*) attacks, and frequently kills, all types of conifer. Usually, the first noticeable symptom is discoloration and withering of the needles, which do not fall. The fungus attacks through the soil, causing white, fan-shaped growths to appear beneath the bark of the roots and trunk at, or just above, soil level. Dark brown root-like structures known as rhizomorphs grow on the diseased roots; these grow out through the soil to spread the disease. Another symptom is the appearance in the fall of honey-colored toadstools at the base of a dying tree. Dig up and burn dead and dying trees together with as many of the roots as possible. Sterilize the soil with a 2 per cent

solution of formalin (1 pt of formalin in 6 gal of water). Apply this at the rate of 6 gal per square yard.

Frost damage may cause die-back of young soft growth in spring, particularly on larches. The needles turn brown but remain attached to the dying shoot tips. Cut off affected shoots to a point an inch or so beyond the apparently dead tissues. Do not plant larches in frost pockets.

Thuja blight (*Didymascella thujina*) is a fungal disease that affects only the western red cedar, *Thuja plicata*, causing individual needles to turn brown. Later, dark brown or black fruiting bodies develop on the upper surface of the needles. After the spores have dispersed the fruiting bodies fall, leaving holes in the foliage. In severe attacks the majority of needles on the lower shoots may be affected, and considerable die-back can occur. The trouble is most serious on hedge plants. There is no fungicide available to amateurs that will control this disease. Instead, cut out affected shoots before the fruiting bodies of the fungus begin their spore production.

Faulty root action due to poor planting, malnutrition or the soil being too wet or too dry can cause conifer shoots to die back. If no definite symptoms of the diseases described above can be seen on a dying plant, check the roots and soil and carry out appropriate remedial measures (see pages 5–7).

Stems galled
Pineapple gall occurs on Norway spruce and is caused by a species of adelgid (*Adelges abietis*). In the spring, adelgid nymphs commence sucking sap from buds. As they do so, they produce chemicals that prevent normal growth of the bud, causing it to swell and grow around the nymphs. This produces a gall that has the appearance of a green pineapple for much of the summer, but during August the galls split open to release the adult adelgids. The gall then dries up and becomes brown, resembling a cone. Little real harm is done to the tree but large numbers of old galls can spoil its appearance. On small trees control the adelgids by spraying with malathion on a mild day, dry day in February or early March.

Stems with visible fungal growth
Rust fungi can affect the shoots of junipers and pines, *Gymnosporangium* spp attacking the former and *Cronartium* spp the latter. On junipers, gelatinous horn-like masses of yellow-orange spores, $\frac{1}{2}$ in or more long, extrude from the swollen parts of the shoots in April to May. On *Pinus sylvestris*, conspicuous yellow blisters bearing spores appear in May and June on swollen parts of the stem. The fungus can girdle branches or even the main stems, causing die-back of the top growth. Similar symptoms appear on five-needled pines such as *Pinus monticola*, *P. strobus* and *P. ayacahuite*, but the swollen parts of branches become spindle-shaped and the blisters grow up to $\frac{1}{2}$ in long and release masses of orange powdery spores. After the spores have been released the blisters dry up, leaving a few shallow scars and exudations of resin. Attacks recur every year until the infected branches die. Cut out affected branches to a point 6–10 in behind the swellings; similar symptoms may, however, occur on other branches the following season. These rusts spend part of their life-cycle on different hosts, thus juniper rusts can also affect *Sorbus* and *Crataegus*, *Pinus sylvestris* rust can affect paeonies and the rust on five-needled pines can infect black currants. Therefore, do not grow these hosts close to conifers.

Fomes root and butt rot (*Fomes annosus*) can affect most conifers, causing die-back and rotting of the heart-wood (butt rot), though affected trees are not always killed. However, the death of the roots can lead to instability which causes the affected tree to fall easily in a wind. The fungus itself is not very conspicuous because the fruiting bodies always appear at ground level and are often obscured by fallen leaves or soil. They are frequently irregular in shape and variable in size, but always have a thin knobbly red-brown upper surface with a white margin, and a white lower surface bearing very small pores. The disease spreads by root contact only and is only likely to occur, therefore, in housing estates that were built on the site of a conifer forest or plantation. Remove and burn any plant affected by this disease since there are no effective chemical controls.

Lawns 1

Introduction

The most common problem on lawns is discolored turf. In most cases this is caused by some adverse cultural condition such as drought, waterlogging, faulty feeding or poor mowing. There are, however, certain pests and diseases that can disfigure, weaken or even kill large areas of turf; for this reason discolored turf should always be investigated early and control measures (if needed) applied as soon as possible.

Seedlings dying

Damping off can be due to species of the fungi *Pythium*, *Fusarium* and *Helminthosporium*. Newly sown seed may rot, and seedlings may die before emerging; if infected after emergence they turn yellow or bronze, and collapse at ground level. Damping off may also show as brown shriveled patches. Prevent infection by using only freshly purchased seed, by treating it with a seed dressing of captan or thiram and by sowing grass seed on a well drained, finely prepared site. Do not sow too thickly or in cold, wet conditions. At the first signs of trouble water with $\frac{1}{2}$ oz of Cheshunt compound in 1 gal of water per square yard.

Grass discolored

Fertilizer scorch is caused by the excessive use of fertilizer or lawn sand, and causes brown or blackened patches, or scorched strips, to appear a few days after feeding. Affected turf should recover in due course, but prevent such troubles by applying fertilizers and lawn sands strictly according to the manufacturer's instructions, and avoid overlapping treated areas. Check the setting of spreaders before use.

Red thread or corticium disease affects fine turf, and is caused by the fungus *Corticium fuciforme*. It usually occurs in late summer and fall, and is most noticeable after rain or dew when pink patches appear on the affected lawn. Horn-like, coral-red growths of fungus spread among the grass and become attached to the stalks and blades, often binding them together. These growths are gelatinous, particularly in humid weather. Although the disease is unsightly, affected plants are rarely killed outright. Sometimes

the trouble only lasts a few days or weeks, in which case only the leaf tips are damaged and no permanent harm is done. In severe attacks the patches persist and, in drier weather, appear bleached. Red thread usually develops where the turf is poorly aerated and the soil is lacking in nitrogen. Prevent attacks by aerating and scarifying the lawn, and by feeding with sulfate of ammonia at $\frac{1}{2}$ oz per square yard during spring and summer (this will also cure slight infections). At the first sign of trouble apply a fungicide or lawn dressing containing benomyl, thiabendazole, thiophanate-methyl or quintozene. Note that the disease will recur shortly after applying a fungicide if nitrogen is not added to the turf.

Faulty root action may be caused by drought, waterlogging, poor aeration (particularly if thatch has built up) or malnutrition. Affected grass becomes yellow or brown and the top growth may die, however, the grass should recover with correct cultural treatment. Prevent such troubles by sowing or laying lawns only on a well drained, finely prepared site, and if necessary improve the drainage. Feed the lawn annually, and aerate and scarify it.

Grass dying in patches

Leatherjackets (*Tipula* spp) are the grubs of crane-flies. They are gray-brown maggots that grow up to $1\frac{1}{2}$ in long and live in the soil where they feed on grass roots. This causes patches of lawn to turn yellow-brown during dry spells in the summer; if these symptoms occur at other times of the year, then they have other causes. Leatherjackets can be detected by watering damaged areas thoroughly and covering them overnight with sacking or black polyethylene. The next day, if leatherjackets are present, they will be found lying on the soil surface beneath the covering. There are no control measures available to amateurs.

Chafer grubs (*Phyllopertha horticola*) cause similar damage to leatherjackets, but are found less frequently in lawns. The grubs have brown heads with creamy-white, C-shaped bodies and three pairs of small legs. When fully grown they are about $\frac{1}{2}$ in long. The adult beetles emerge from the soil at dusk in late May or June. There are no control measures available against this pest.

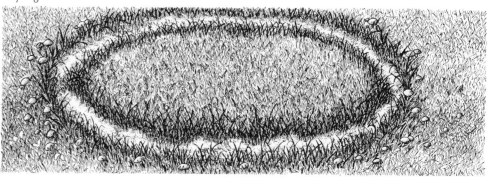

Fairy ring

Chafer grub

Leatherjackets

Snow mold, or fusarium patch, is caused by the fungus *Fusarium nivale* and can occur at any time of the year if the weather is moist, but the disease is most severe in October and during mild spells in the winter and spring. As the name suggests, it may also occur after snow has thawed, particularly where compaction has occurred. The disease first shows as small patches of yellow dying grass, which later turn brown, increase in size and coalesce. In moist weather the patches become covered with a white, or faintly pink, cotton-like growth of fungus which mats the dying vegetation together. The disease is encouraged by lack of aeration and excessive use of nitrogenous fertilizers, especially when applied after August. Prevent infection by good cultural treatment and by scarifying the turf to improve aeration. In September aerate to a depth of 3 in or more using, if possible, a hollow-tine tool. Disperse heavy morning dew with a besom or

flexible bamboo cane to keep the grass as dry as possible. Mow regularly and remove overhanging vegetation to encourage a better circulation of air. If possible, keep off a snow-covered lawn. To control mild attacks and check more severe infection, water with a solution of $\frac{1}{4}$ oz of iron sulfate in $\frac{1}{2}$ gal of water per square yard. For more virulent attacks apply a lawn dressing or a fungicide containing benomyl, thiabendazole, thiophanate-methyl or quintozene.

Circular patches present

Dollar spot (*Sclerotinia homoeocarpa*) is usually found during humid weather in late summer. It appears as small straw-colored circular patches, 1–2 in in diameter which may coalesce. To control dollar spot, apply those fungicides recommended for snow mold. In addition feed the lawn with nitrate of ammonia during the spring and summer to encourage vigor.

Lawns 2

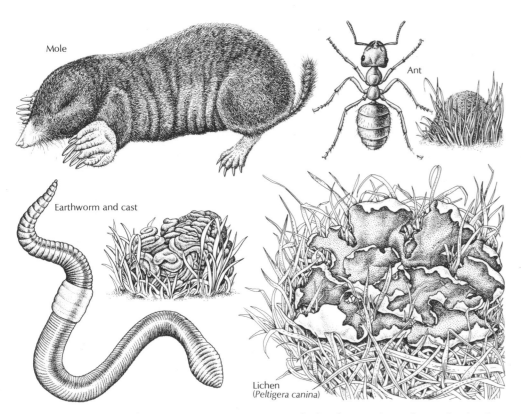

Mole

Ant

Earthworm and cast

Lichen
(*Peltigera canina*)

Dark green rings in turf
Fairy rings are caused by various fungi, including species of *Lycoperdon* (puffballs) and *Agaricus* (mushrooms), and *Marasmius oreades*, which is the commonest and most troublesome. The threads of these fungi grow in the soil between or beneath the roots of turf grasses and spread outwards in ever widening circles. One or more zones of lush green grass, which may be continuous or interrupted, develop at the periphery of the ring. Species of *Lycoperdon* and *Agaricus* produce one ring of dark green grass with few fungal threads in the soil. *M. oreades* produces a ring composed of an outer and inner zone of vigorous dark green grass with an area between them where the grass is brown and dead or the soil is bare. Slender toadstools $1\frac{1}{2}$–4 in high with brown caps 1–$2\frac{1}{2}$ in across appear in summer and fall, especially in wet weather, at the outer edge of the dead zone. A dense mat of fungal

growth develops in the soil to a depth of at least 9 in and, since it is impenetrable by water, the grass dies through drought as well as starvation caused by competition from the fungus. Single rings can usually be controlled by watering with a solution of 1 lb of iron sulfate in 3 gal of water applied at $\frac{1}{2}$ gal per square yard—ensure that the affected area is first soaked thoroughly with water. Established rings caused by *M. oreades* are difficult to control. Lift affected turf for 1 ft beyond the inner and outer rings and remove it from the site and burn it, taking care not to spill any infected soil. Fork the exposed soil strip to 9 in, breaking it up finely. Where threads are deeper, make crowbar holes at 6 in intervals in the dug area. Water in a solution of 2 per cent formalin plus a wetting agent, applying 1 gal per square yard of soil. Make sure that no formalin is spilled on the adjacent grass. Cover with damp sacks or sheet polyethylene

and leave for seven to ten days, then fork it over and leave for two weeks if the soil is light or up to five weeks if it is heavy. Finally, fill with fresh soil and re-seed or re-sod.

Visible growths
Other toadstools, apart from those causing fairy rings (see above), commonly appear on lawns during the autumn. These are harmless and need no specific control measures apart from brushing them up if they are too unsightly, but they are generally short-lived and disappear very quickly of their own accord. They can, however, be suppressed by watering the affected areas with a solution of 1 oz of iron sulfate in 3 gal of water applied at 1 gal per square yard. Some toadstools develop year after year if there is a lot of woody debris beneath the soil.

Lichens such as *Peltigera canina* are composite organisms, each being a combination of a fungus and an alga. They consist of overlapping leaf-like structures growing horizontally in the turf. *P. canina* is variable in color, being deep green-black when moist or gray-green or brown in dry weather. The upper surface has a leathery texture whereas the lower surface is white and spongy. Lichens flourish in poorly drained soil, but may occur on well drained turf if the surface is compacted or growing in shaded areas under trees where the soil is impoverished. Rake out the growths and treat the affected areas with iron sulfate at $\frac{1}{2}$ oz per square yard bulked with 4 oz of sand. The growths will, however, reappear unless the lawn is aerated and fed, and, if necessary, the drainage is improved.

Slime molds are organisms that are intermediate between bacteria and fungi. They usually occur in late spring or early fall following heavy rain, and, although they appear to smother grass, are purely superficial, though unsightly. These growths may be white or yellow, and produce small gray fruiting bodies which subsequently release masses of purple-brown spores. Once these spores have been released the growths usually disappear fairly rapidly of their own accord. The process may be hastened by washing the spore masses away from the lawn with a stream of water.

Lawn surface slippery
Algae and lichens of various types may create a slippery gelatinous layer over the surface of the lawn. They are usually dark green or black, and appear on turf where the surface is perpetually damp from poor drainage, compaction or the dripping of overhanging trees. Treat affected areas with either 1 oz of copper sulfate in 30 gal of water for every 100 sq yd or a proprietary product containing dichlorophen. To prevent recurrence of the growths check for blocked drains or compaction, and drain or spike accordingly.

Soil heaps present
Moles (*Talpa europaea*) build extensive underground tunnel systems, depositing the excavated soil in heaps on the surface. Outside the breeding season, which is during the early spring, moles tend to live solitary lives, with each having its own tunnel system. Thus all the molehills in a small garden could be the work of a single mole. Little harm is done to the lawn by tunneling but molehills are a nuisance since they interfere with mowing. Mole traps are available from garden and gun shops, and these are the most effective means of eliminating moles. Mole smokes sometimes kill the mole, but more often it is only driven away temporarily.

Earthworms (*Allolobophora* spp) ingest soil and organic matter as they burrow through the earth. They excrete small heaps of soil (known as worm casts) on to the lawn surface. These can make the lawn slippery and, if they are smeared by trampling or mowing, create bare patches that may be colonized by weeds. Worm activity is at its greatest in spring and fall when the soil is warm and moist. These are the most effective times to take control measures since worms then live close to the soil surface. Do this by watering the lawn with potassium permanganate or derris. These bring worms to the surface where they can be raked up.

Ants (various species) nest in the soil and, during the summer, are liable to throw up small heaps of fine soil particles. These may become smeared by the mower to provide sites where weeds can establish. There is no satisfactory control against ants.

Index 1

Index 2/Acknowledgements

The Royal Horticultural Society and the Publishers can accept no liability either for failure to control pests and diseases, and to rectify disorders, by the methods recommended, or for any consequences of these methods. We specifically draw the reader's attention to the necessity of carefully reading and following the manufacturer's instructions on any product.

Acknowledgements

Most of the artwork in this book has been based on photographs from the Royal Horticultural Society's collection. Further references were kindly supplied by East Malling Research Station, Brian Furner, Long Ashton Research Station, Ministry of Agriculture Fisheries and Food, The Scottish Horticultural Research Institute and Shell International Petroleum Co. Ltd.

Artists: Lindsay Blow, Charles Chambers, William Giles, Tony Graham, Edwina Keene, Sandra Pond, Ed Roberts, Paul Stafford, Lorna Turpin, John Woodcock.